#11 Shizuka Hoshijiro's Return

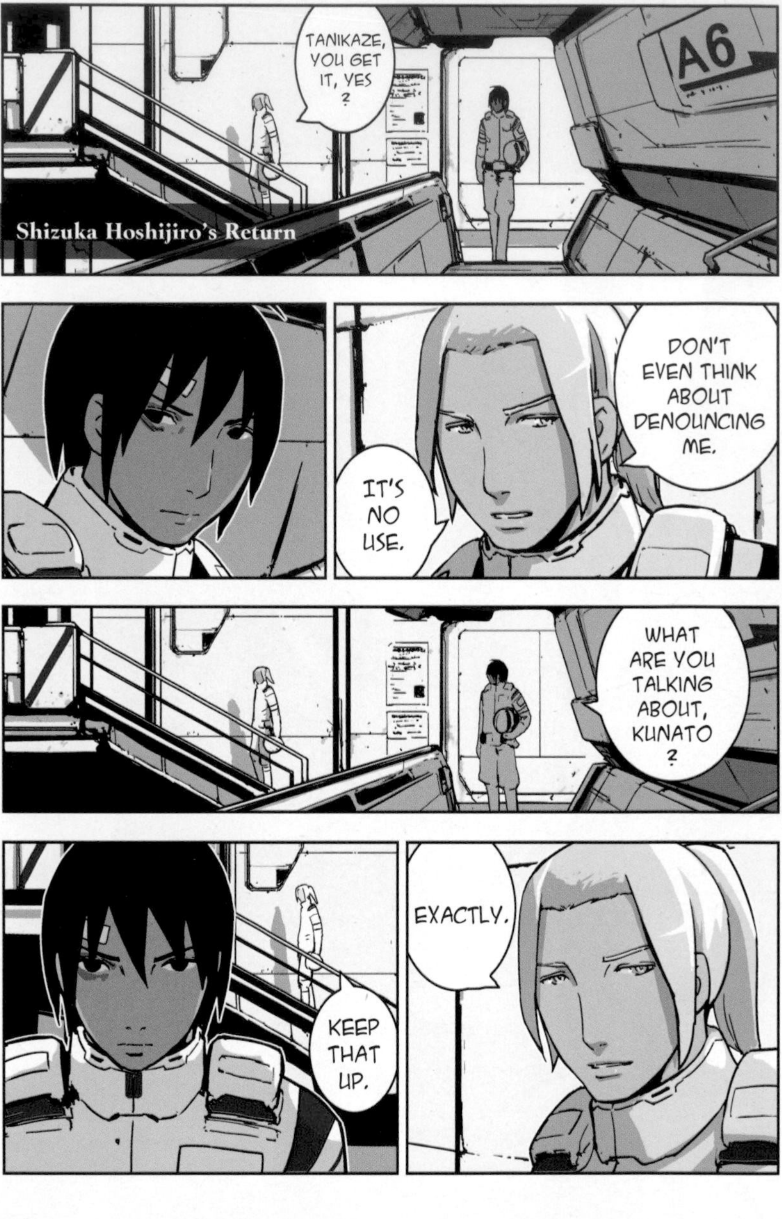
TANIKAZE, YOU GET IT, YES?
DON'T EVEN THINK ABOUT DENOUNCING ME.
IT'S NO USE.
WHAT ARE YOU TALKING ABOUT, KUNATO?
EXACTLY.
KEEP THAT UP.

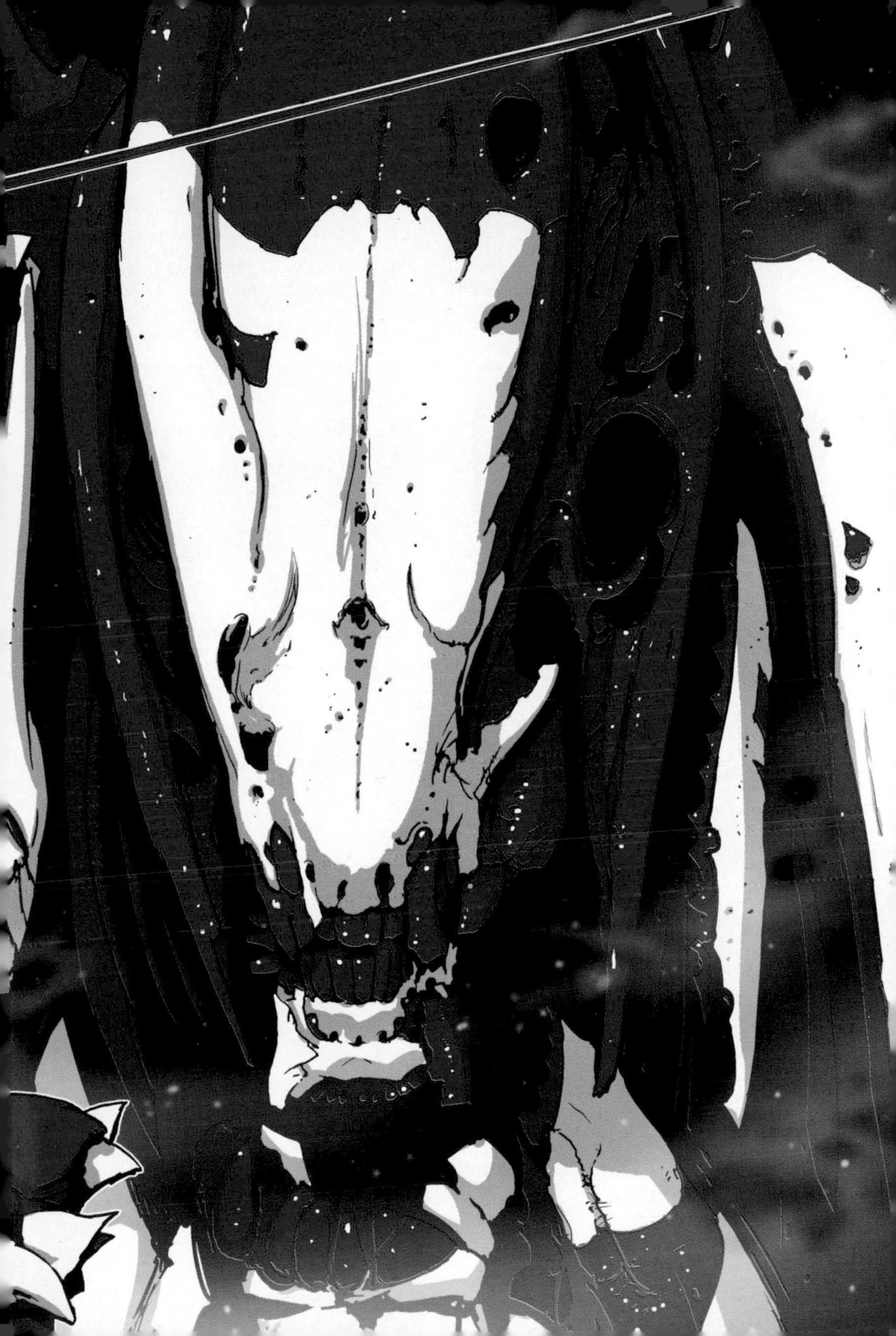

KNIGHTS OF SIDONIA ③

TSUTOMU NIHEI

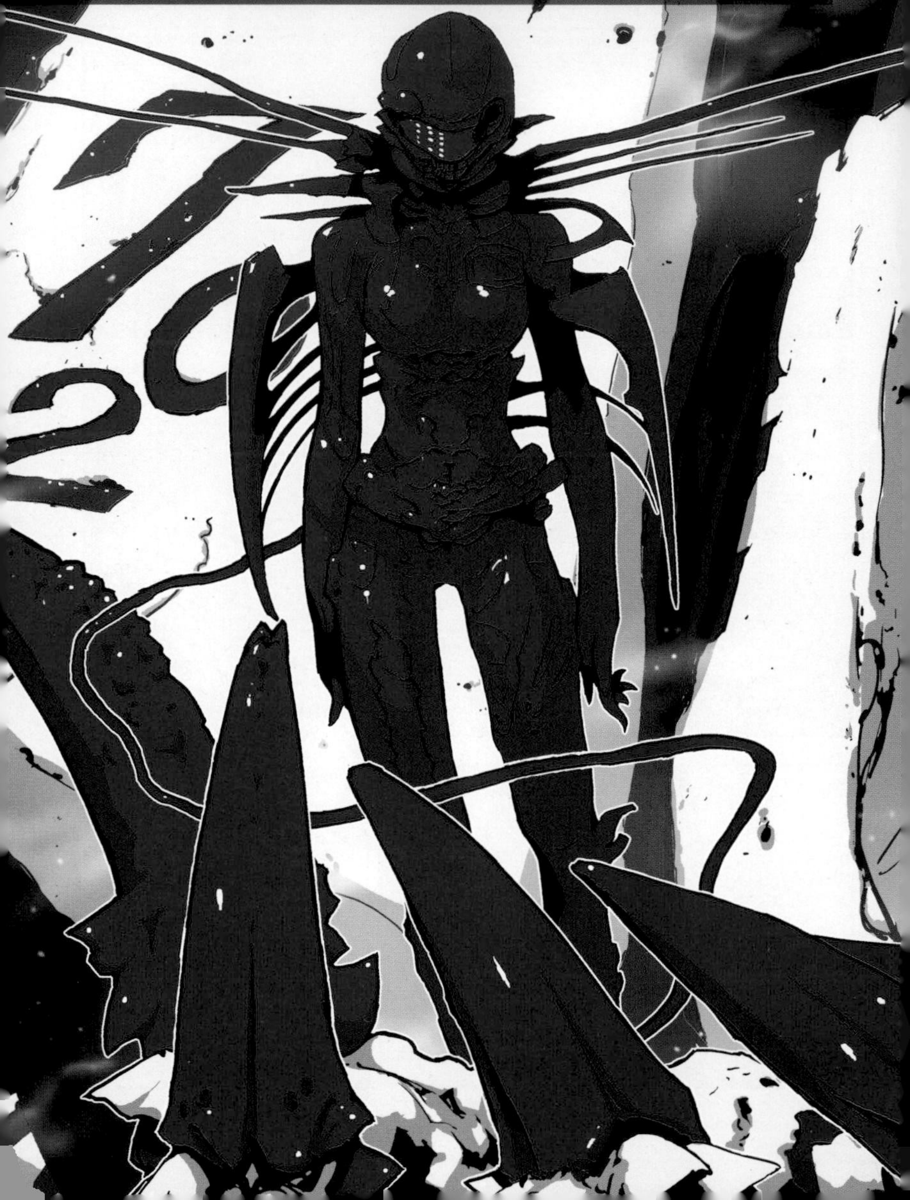

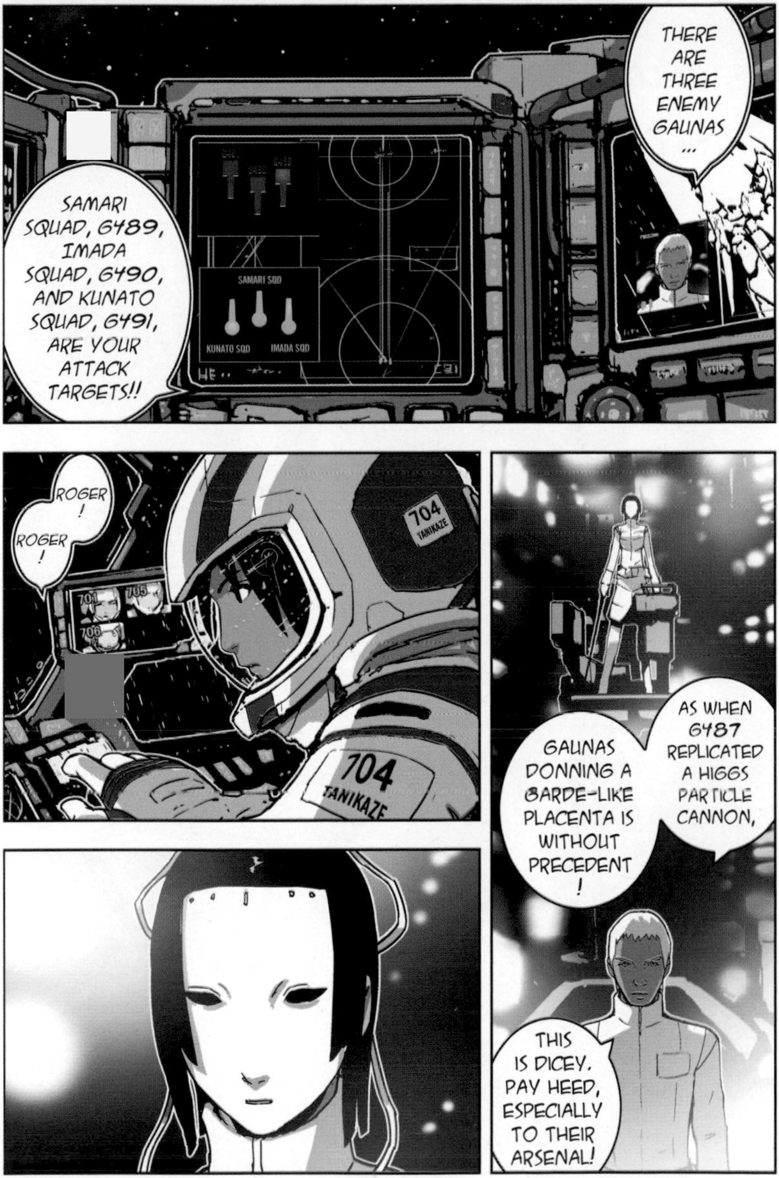
THERE ARE THREE ENEMY GAUNAS ...
SAMARI SQUAD, 6489, IMADA SQUAD, 6490, AND KUNATO SQUAD, 6491, ARE YOUR ATTACK TARGETS!!
SAMARI SQD
KUNATO SQD
IMADA SQD
AS WHEN 6487 REPLICATED A HIGGS PARTICLE CANNON,
GAUNAS DONNING A GARDE-LIKE PLACENTA IS WITHOUT PRECEDENT!
THIS IS DICEY. PAY HEED, ESPECIALLY TO THEIR ARSENAL!
ROGER!
ROGER!
704 TANIKAZE
704 TANIKAZE

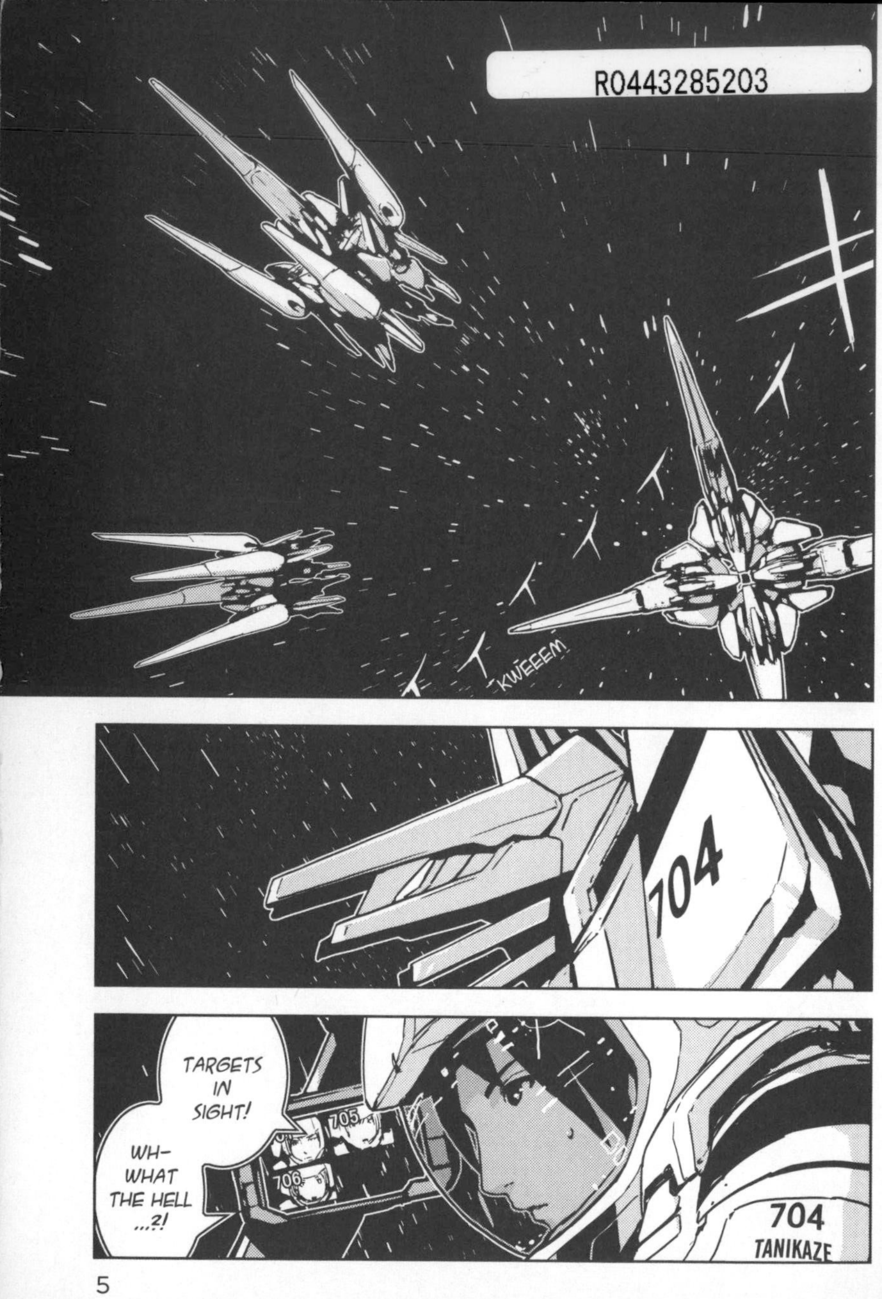
KWEEEM
704
704
TANIKAZE
705
706
TARGETS IN SIGHT!
WH- WHAT THE HELL ...?!

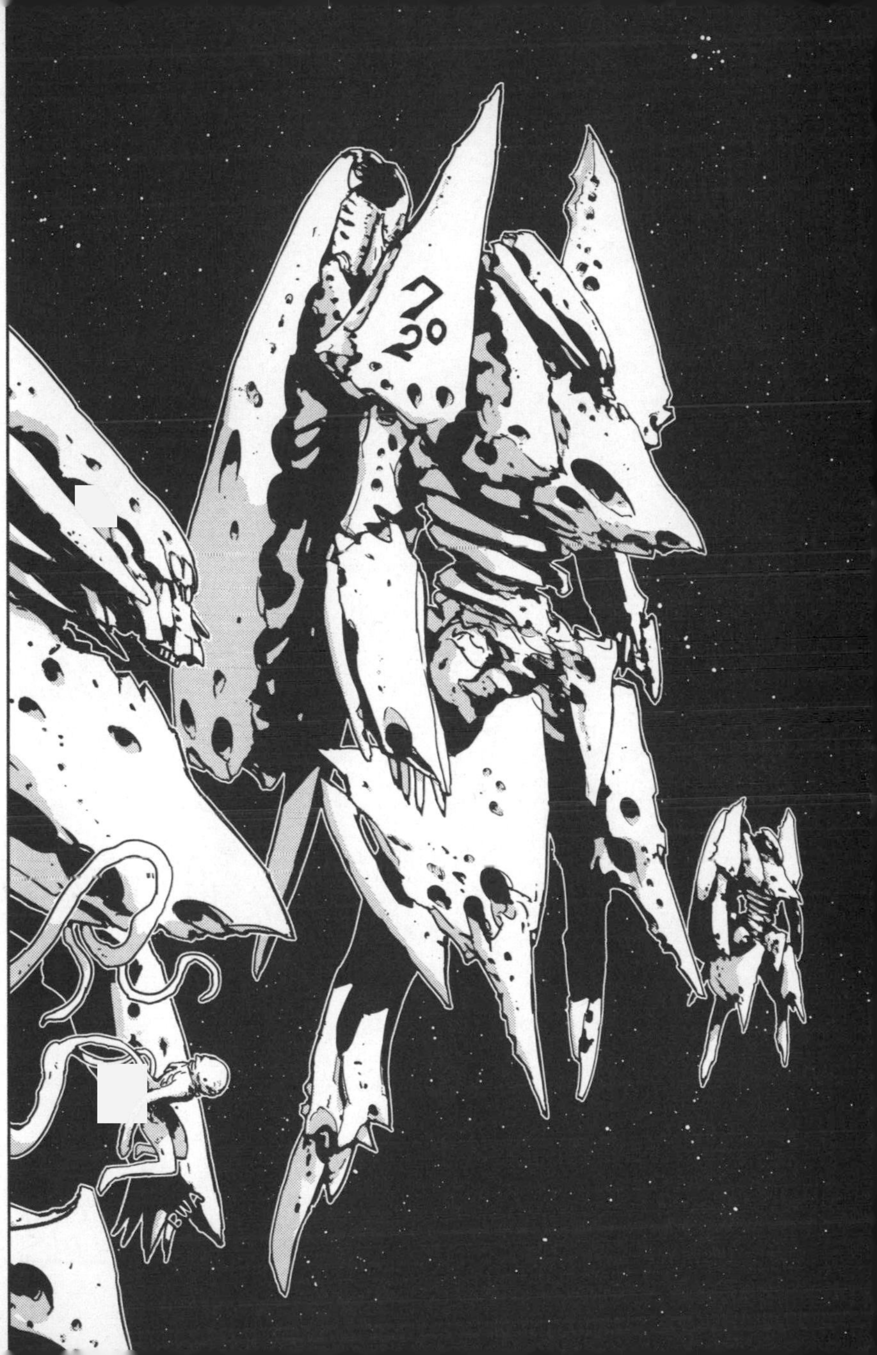
7
20
BWA

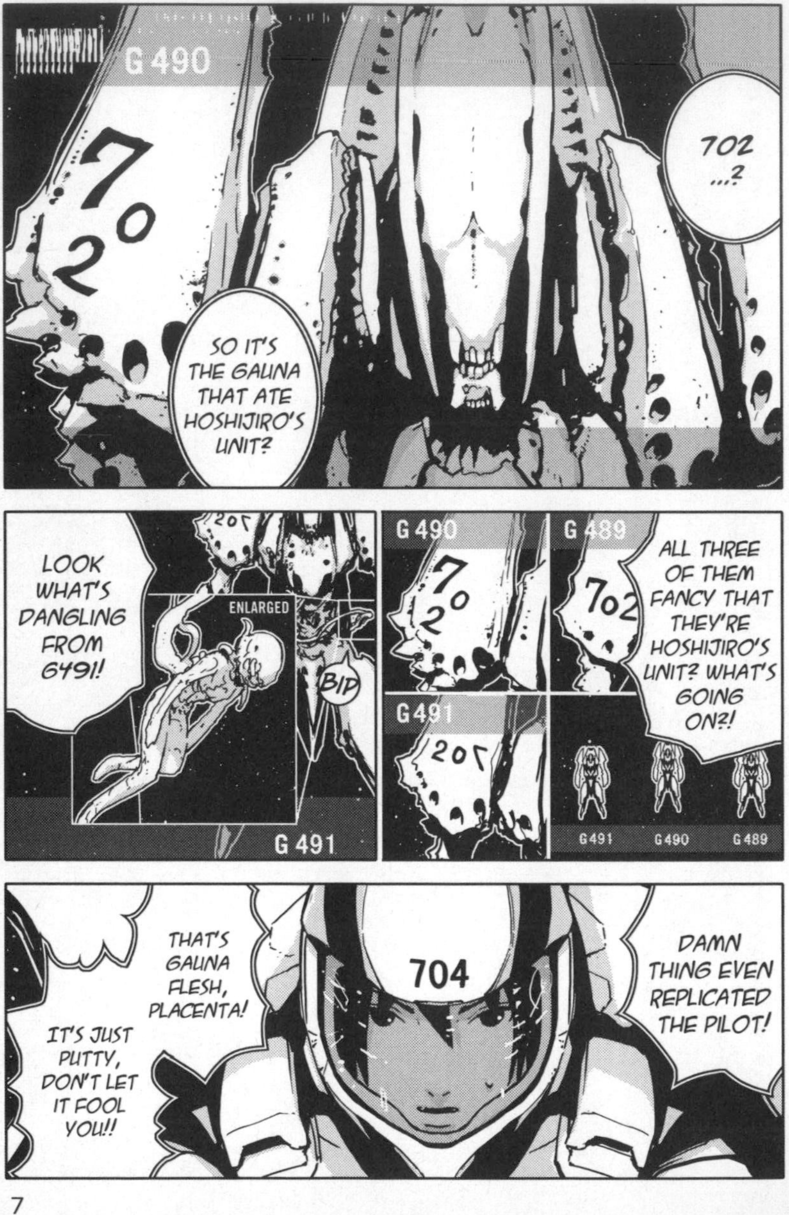

702 ...?
G 490
702
SO IT'S THE GAUNA THAT ATE HOSHIJIRO'S UNIT?
G 489
702
ALL THREE OF THEM FANCY THAT THEY'RE HOSHIJIRO'S UNIT? WHAT'S GOING ON?!
G 490
702
G 491
702
G491 G490 G489
LOOK WHAT'S DANGLING FROM G491!
ENLARGED
BIP
G 491
DAMN THING EVEN REPLICATED THE PILOT!
704
THAT'S GAUNA FLESH, PLACENTA!
IT'S JUST PUTTY, DON'T LET IT FOOL YOU!!

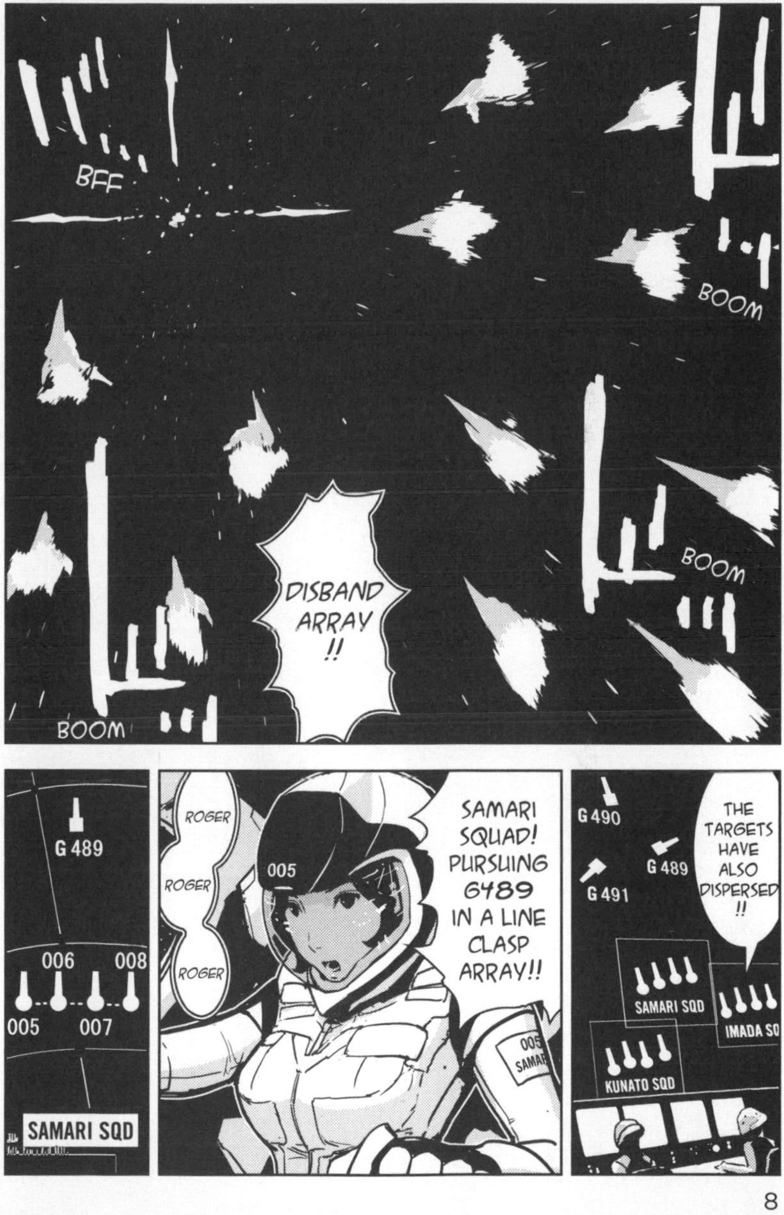
BOOM
BFF
BOOM
DISBAND ARRAY!!
BOOM
THE TARGETS HAVE ALSO DISPERSED!!
G 490
G 489
G 491
SAMARI SQD
IMADA SQ
KUNATO SQD
SAMARI SQUAD! PURSUING G489 IN A LINE CLASP ARRAY!!
005
ROGER
ROGER
ROGER
G 489
006
008
005
007
SAMARI SQD

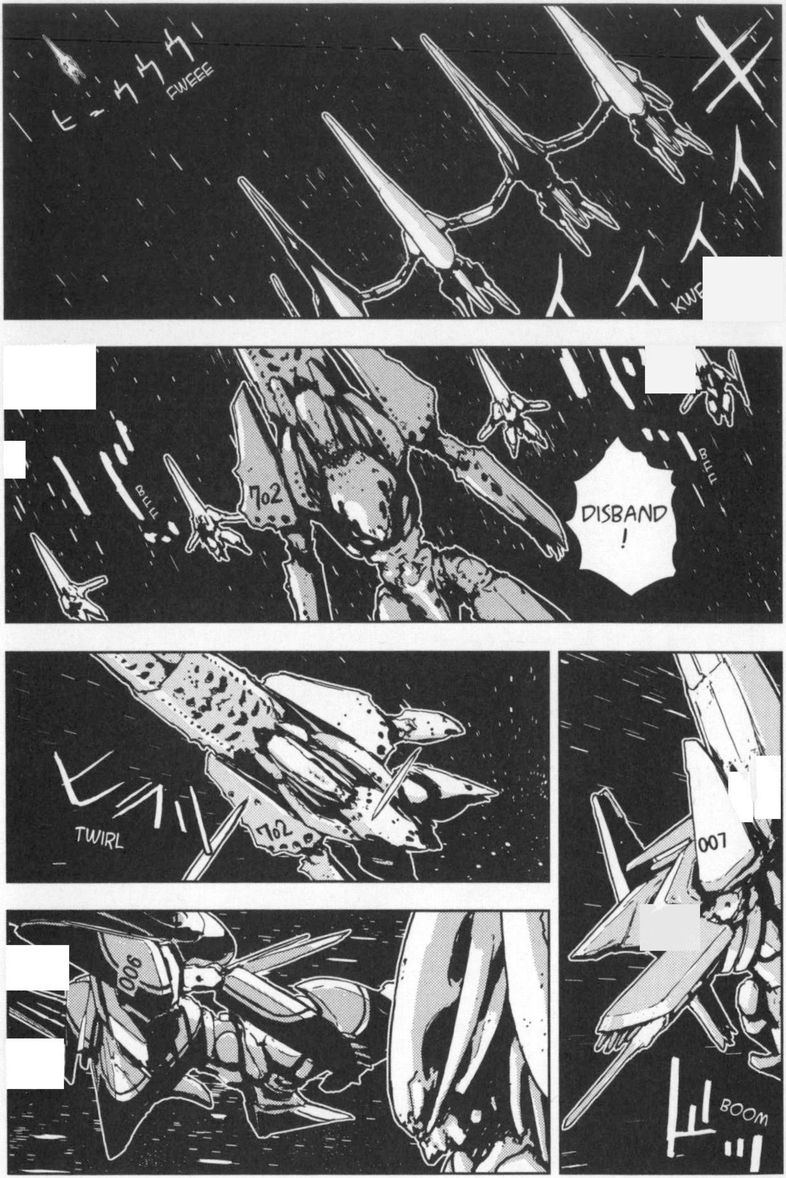
FWEEE
DISBAND
!
BOOM
TWIRL

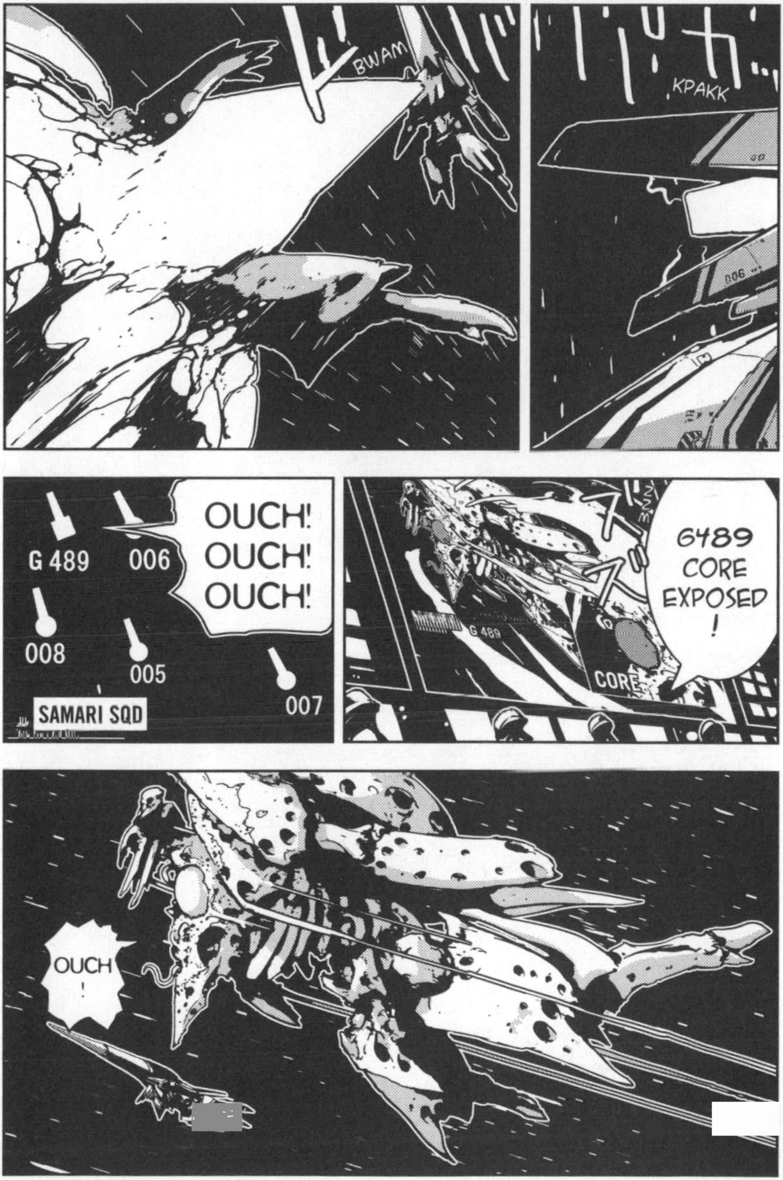
KPAKK
006
BWAM
6489 CORE EXPOSED !
G 489
CORE
OUCH! OUCH! OUCH!
G 489
006
008
005
007
SAMARI SQD
OUCH !

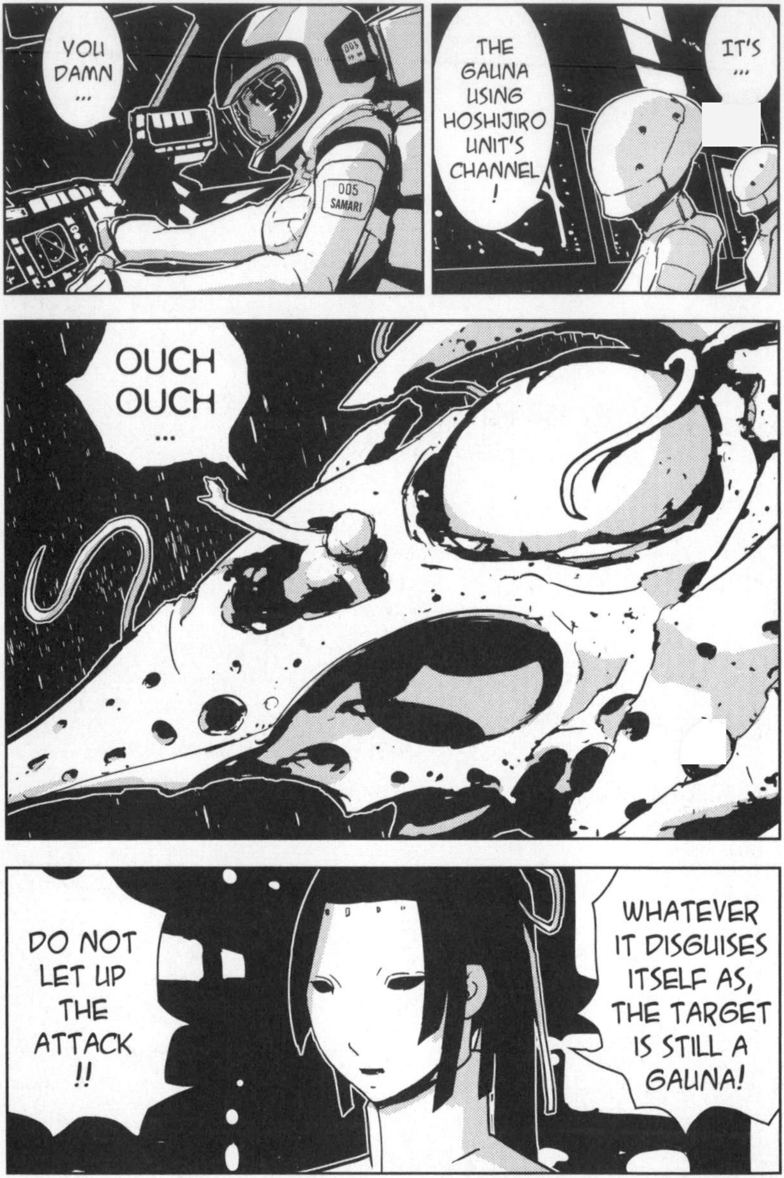
IT'S ...
THE GAUNA USING HOSHIJIRO UNIT'S CHANNEL !
YOU DAMN ...
005 SAMARI
OUCH OUCH ...
WHATEVER IT DISGUISES ITSELF AS, THE TARGET IS STILL A GAUNA!
DO NOT LET UP THE ATTACK !!

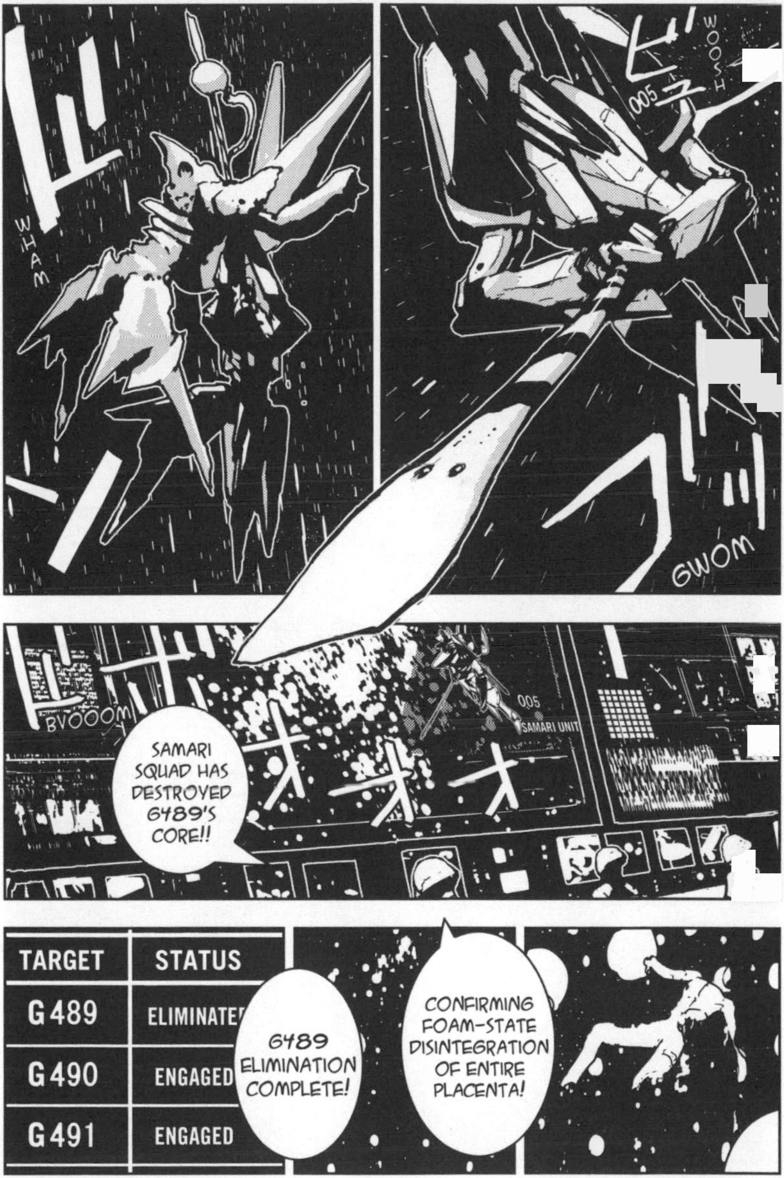
WOOSH
005
GWOM
WHAM
BVOOOM
005
SAMARI UNIT
SAMARI SQUAD HAS DESTROYED 6489'S CORE!!
CONFIRMING FOAM-STATE DISINTEGRATION OF ENTIRE PLACENTA!
6489 ELIMINATION COMPLETE!
TARGET STATUS
G489 ELIMINATE
G490 ENGAGED
G491 ENGAGED

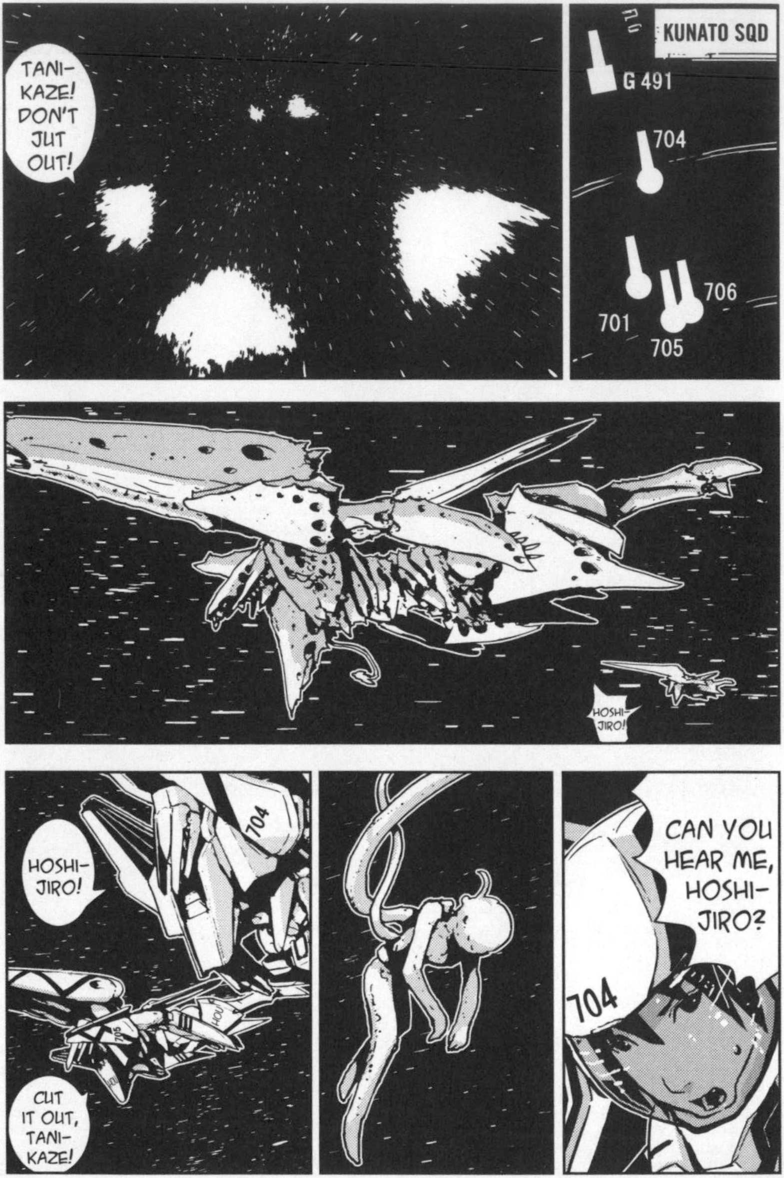
KUNATO SQD
G 491
704
706
701
705
TANI-KAZE! DON'T JUT OUT!
HOSHI-JIRO!
CAN YOU HEAR ME, HOSHI-JIRO?
704
704
HOSHI-JIRO!
CUT IT OUT, TANI-KAZE!

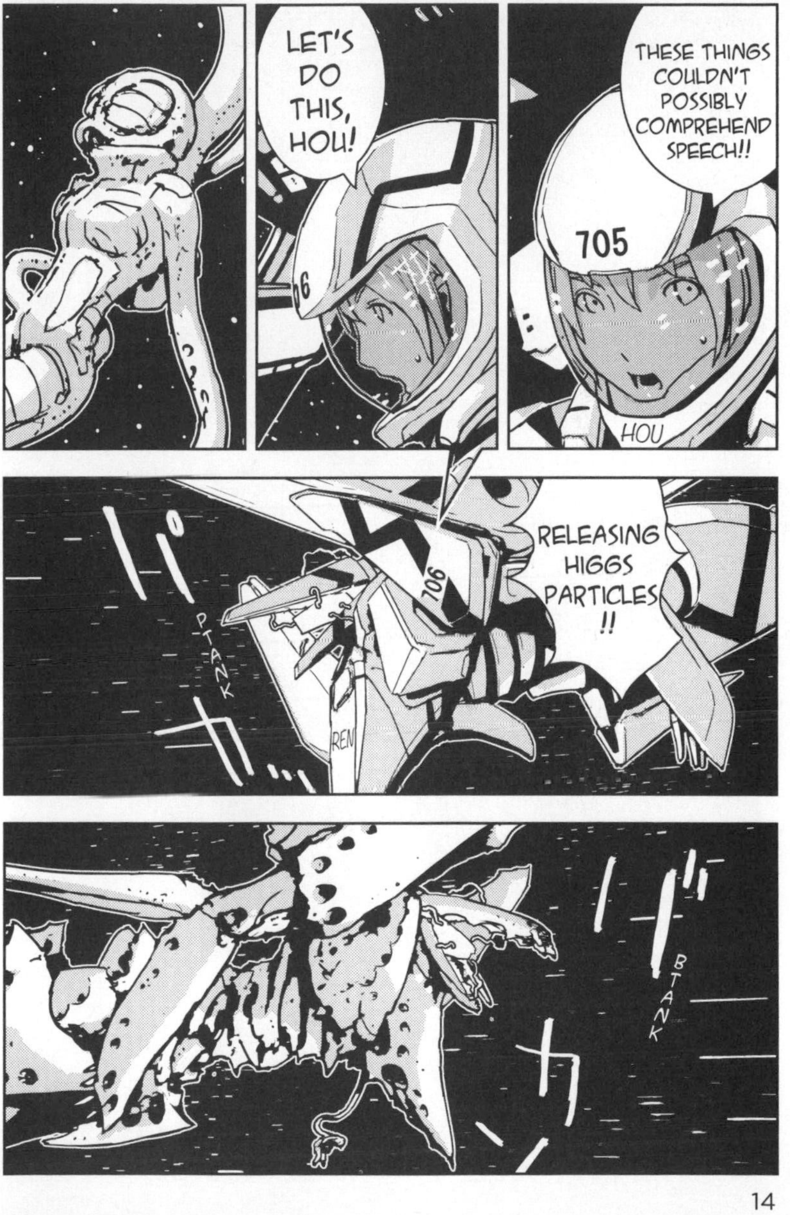
THESE THINGS COULDN'T POSSIBLY COMPREHEND SPEECH!!
705
HOU
LET'S DO THIS, HOU!
RELEASING HIGGS PARTICLES!!
706
PTANK
REN
BTANK

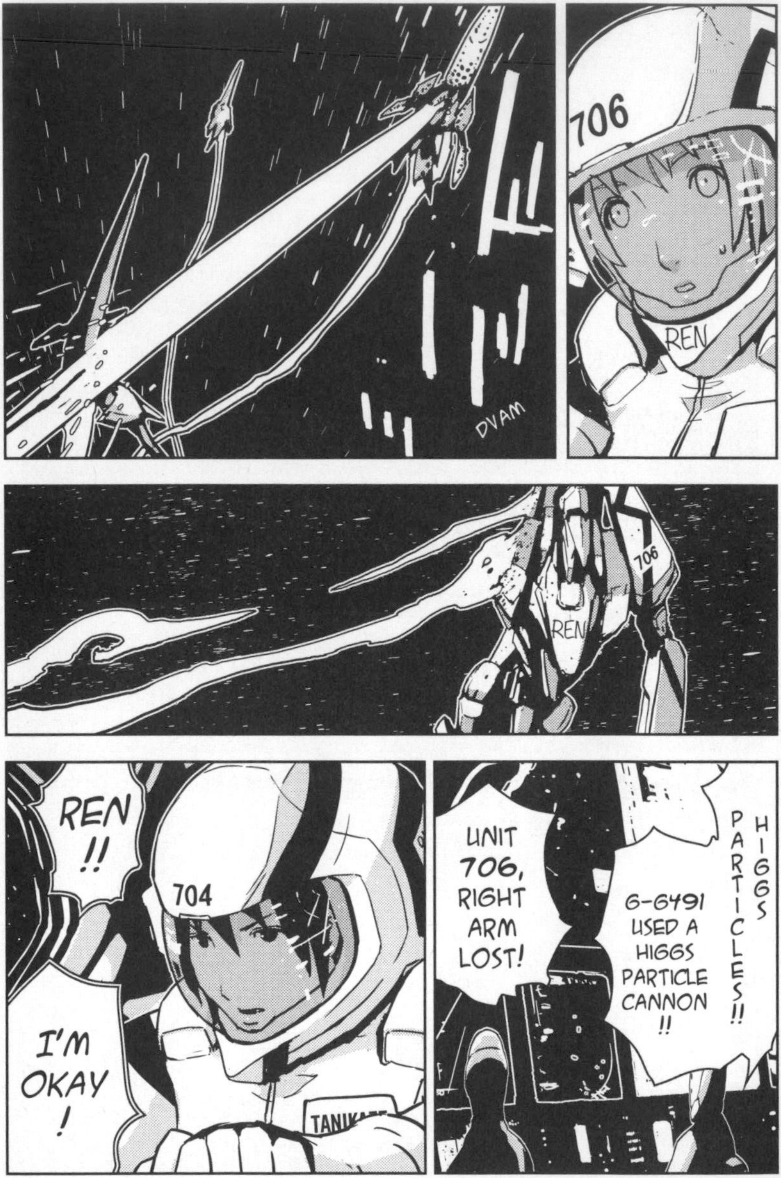

706
REN
DVAM
706
REN
HIGGS PARTICLES!!
6-6491 USED A HIGGS PARTICLE CANNON!!
UNIT 706, RIGHT ARM LOST!
REN!!
704
I'M OKAY!

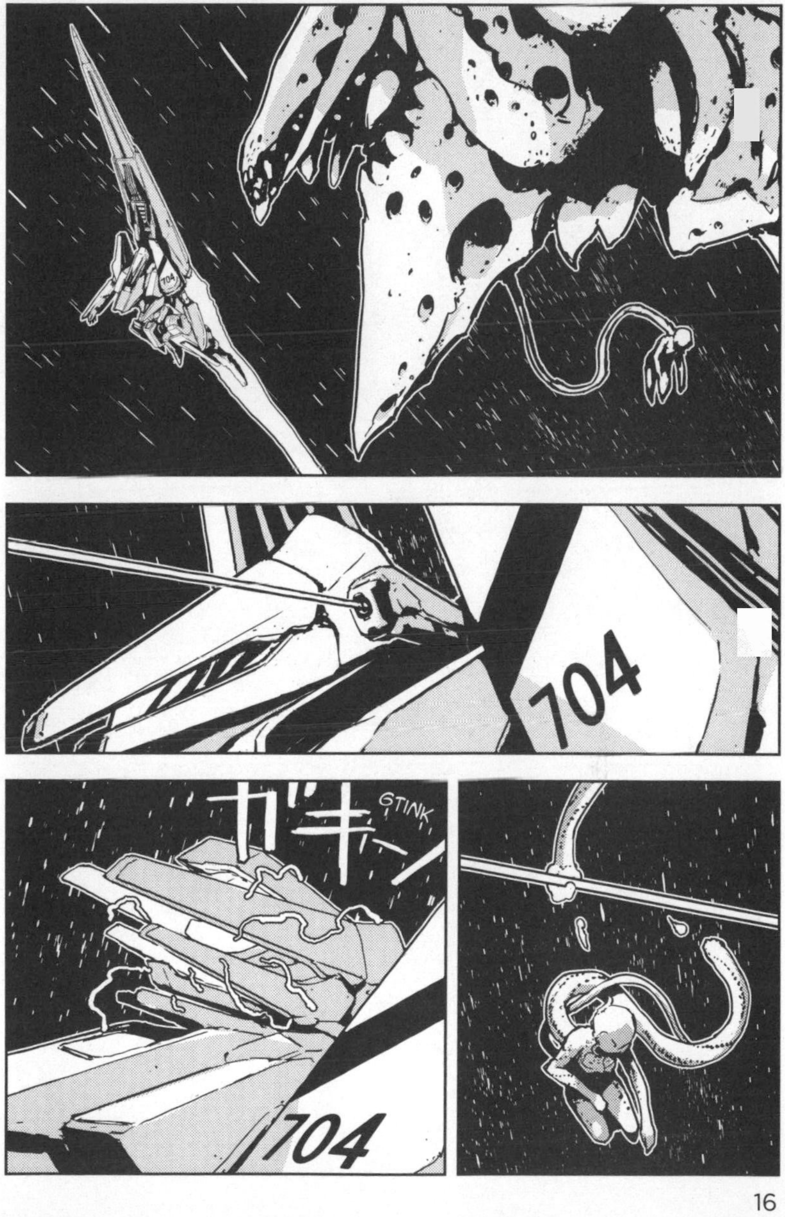
704
704
ガキーン
GTINK
704

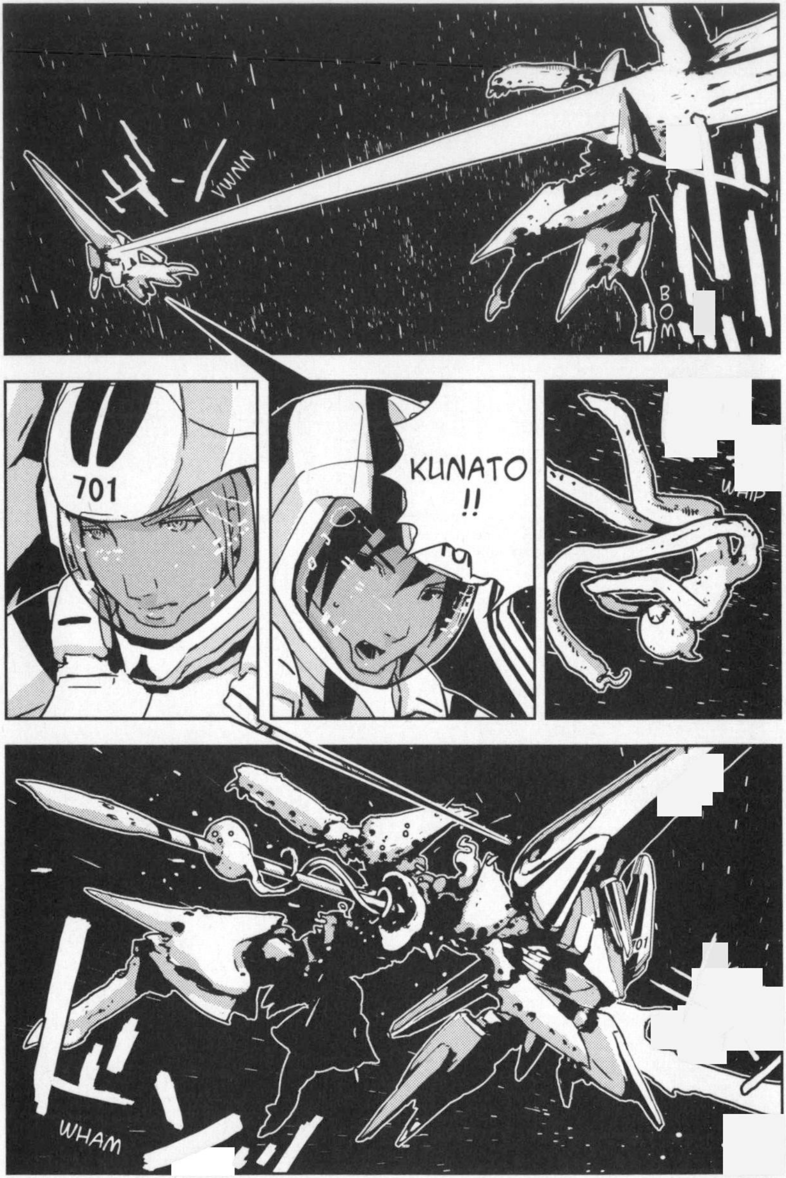
BOM
VWNN
WHIP
KUNATO !!
WHAM

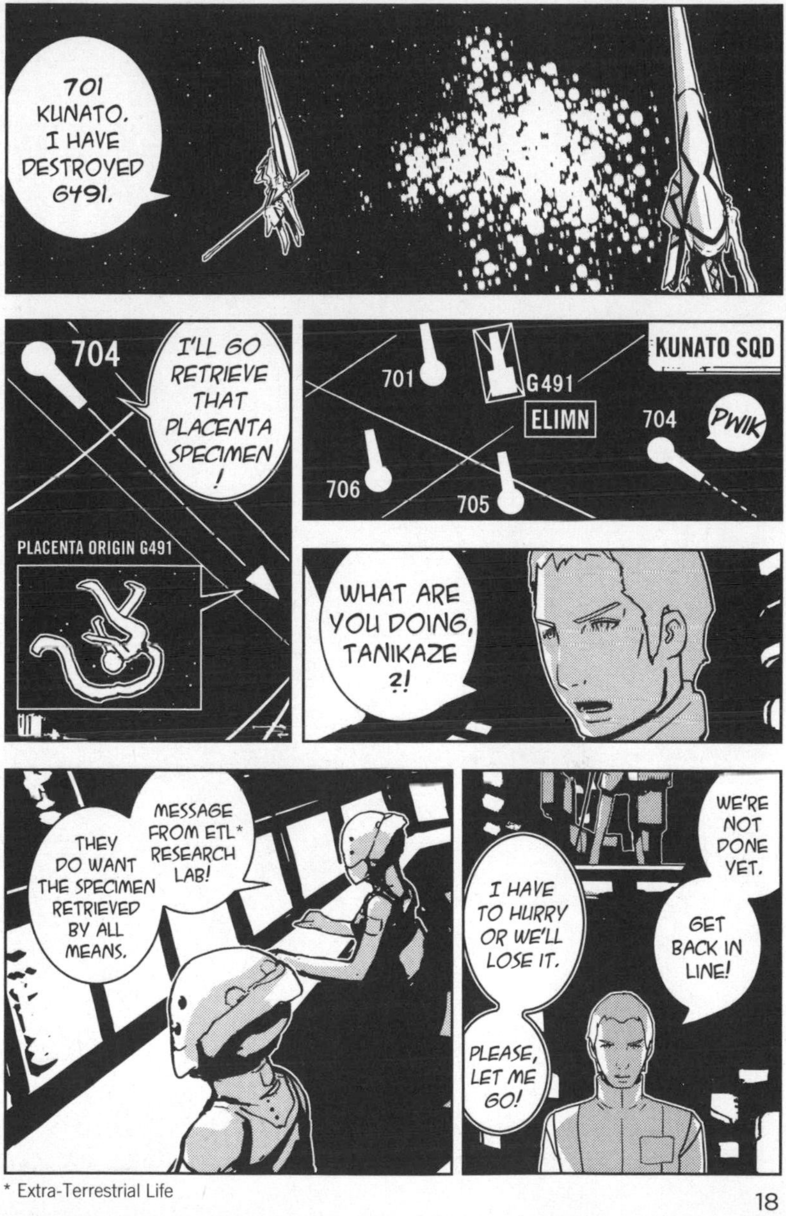
701 KUNATO. I HAVE DESTROYED G491.
KUNATO SQD
701
G491
ELIMN
704
PWIK
706
705
WHAT ARE YOU DOING, TANIKAZE ?!
704
I'LL GO RETRIEVE THAT PLACENTA SPECIMEN !
PLACENTA ORIGIN G491
WE'RE NOT DONE YET.
GET BACK IN LINE!
I HAVE TO HURRY OR WE'LL LOSE IT.
PLEASE, LET ME GO!
MESSAGE FROM ETL* RESEARCH LAB!
THEY DO WANT THE SPECIMEN RETRIEVED BY ALL MEANS.

* Extra-Terrestrial Life

RETRIEVAL PERMITTED.
KUNATO SQUAD, EXCLUDING TANIKAZE, AND SAMARI SQUAD, GO ASSIST IMADA SQUAD!
ROGER!
ROGER!
TANIKAZE, CONTINUE RETRIEVAL OF SPECIMEN!
ROG-ER!!
THAT WAY!!
IMADA SQD
139
312
G 490
217
011
TARGET | STATUS
G489 | ELIMINATED
G490 | ENGAGED
G491 | ELIMINATED
GWOM
GAH!
FOOM
BWWM
REST-LESS, AREN'T YA?
DAMN...
139 IMADA
IT'S READING EACH OF OUR MOVES.

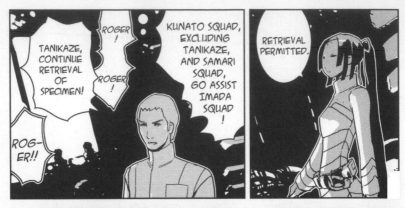

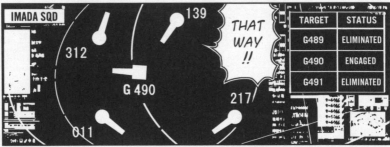

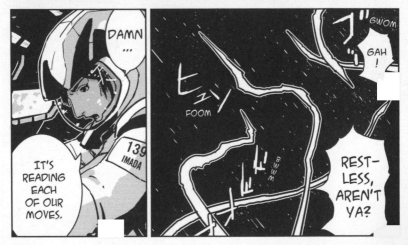

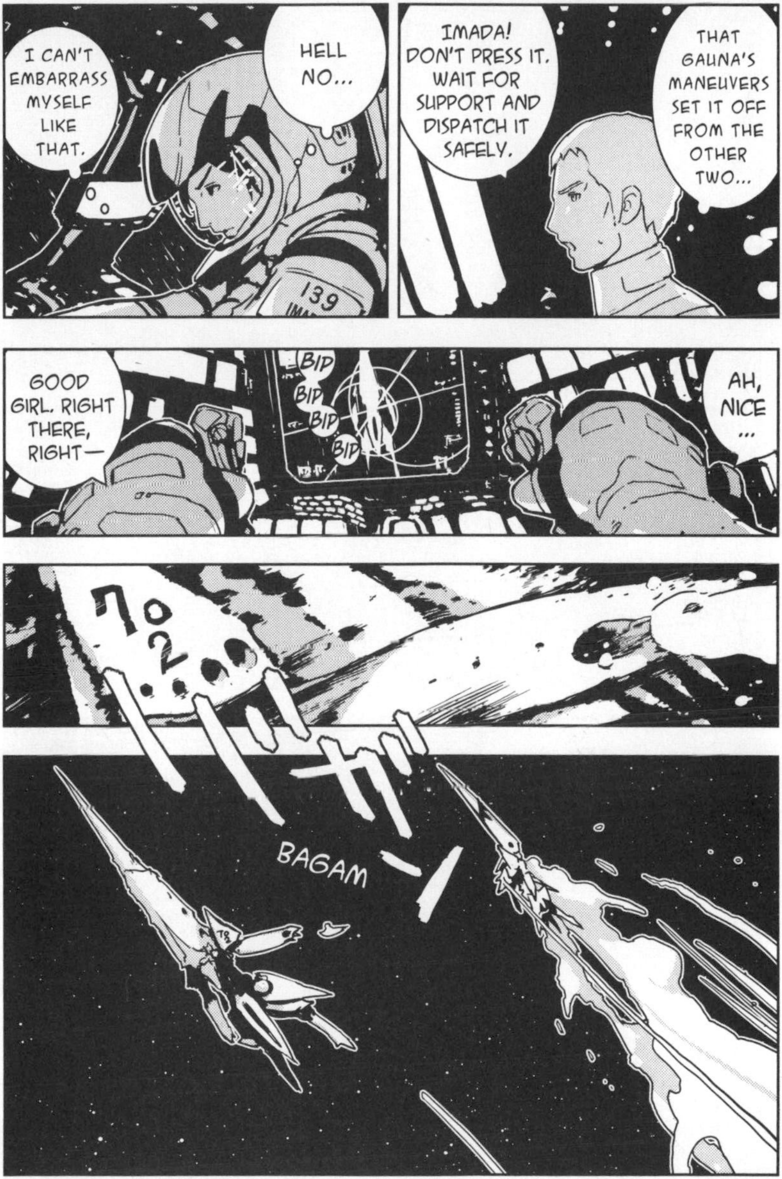
THAT GAUNA'S MANEUVERS SET IT OFF FROM THE OTHER TWO...
IMADA! DON'T PRESS IT. WAIT FOR SUPPORT AND DISPATCH IT SAFELY.
HELL NO...
I CAN'T EMBARRASS MYSELF LIKE THAT.
139
AH, NICE ...
BIP
BIP
BIP
BIP
GOOD GIRL. RIGHT THERE, RIGHT—
702
BAGAM

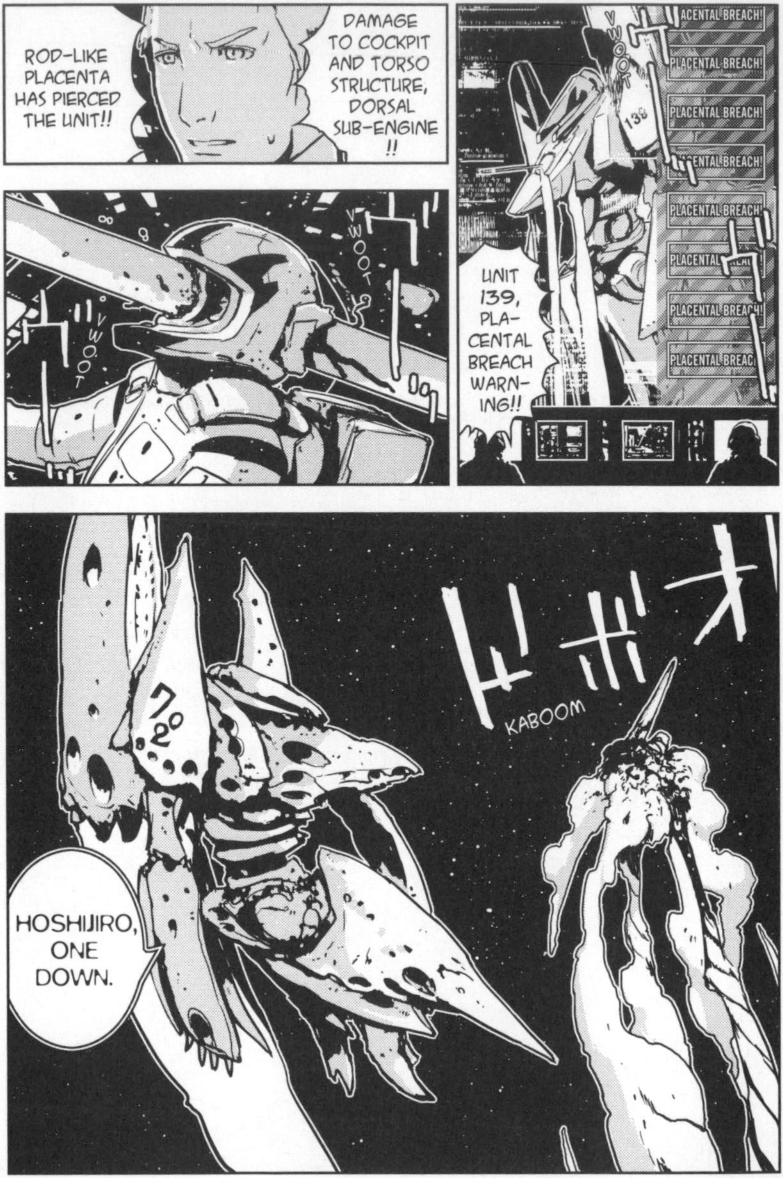
ACENTAL BREACH!
PLACENTAL BREACH!
PLACENTAL BREACH!
VWOOT
138
PLACENTAL BREACH!
PLACENTAL BREACH!
UNIT 139, PLACENTAL BREACH WARNING!!
DAMAGE TO COCKPIT AND TORSO STRUCTURE, DORSAL SUB-ENGINE !!
ROD-LIKE PLACENTA HAS PIERCED THE UNIT!!
VWOOT
VWOOT
KABOOM
HOSHIJIRO, ONE DOWN.

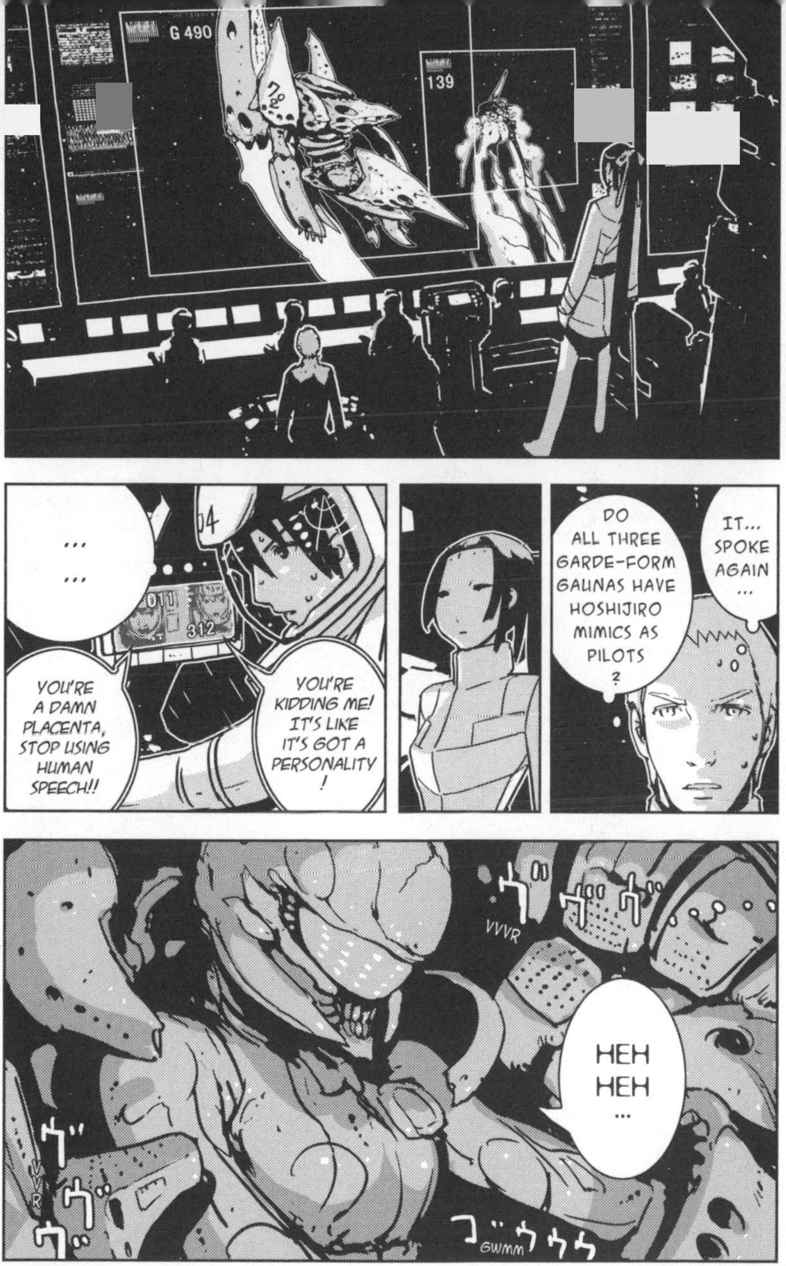
G 490
139
IT... SPOKE AGAIN...
DO ALL THREE GARDE-FORM GAUNAS HAVE HOSHIJIRO MIMICS AS PILOTS?
... ...
YOU'RE KIDDING ME! IT'S LIKE IT'S GOT A PERSONALITY!
YOU'RE A DAMN PLACENTA, STOP USING HUMAN SPEECH!!
VVVR
HEH HEH ...
VVR
GWMM

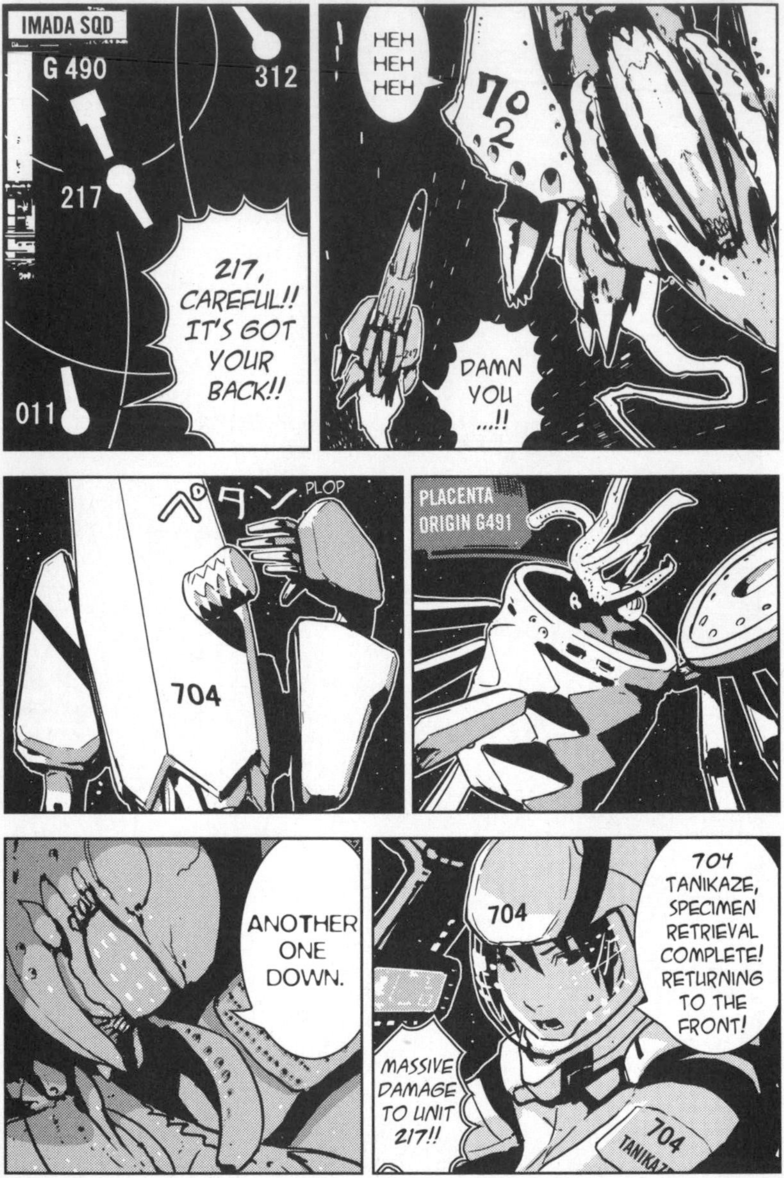
HEH HEH HEH
702
DAMN YOU ...!!
IMADA SQD
G 490
312
217
217, CAREFUL!! IT'S GOT YOUR BACK!!
011
PLACENTA ORIGIN G491
ペタン PLOP
704
704 TANIKAZE, SPECIMEN RETRIEVAL COMPLETE! RETURNING TO THE FRONT!
704
MASSIVE DAMAGE TO UNIT 217!!
704 TANIKAZE
ANOTHER ONE DOWN.

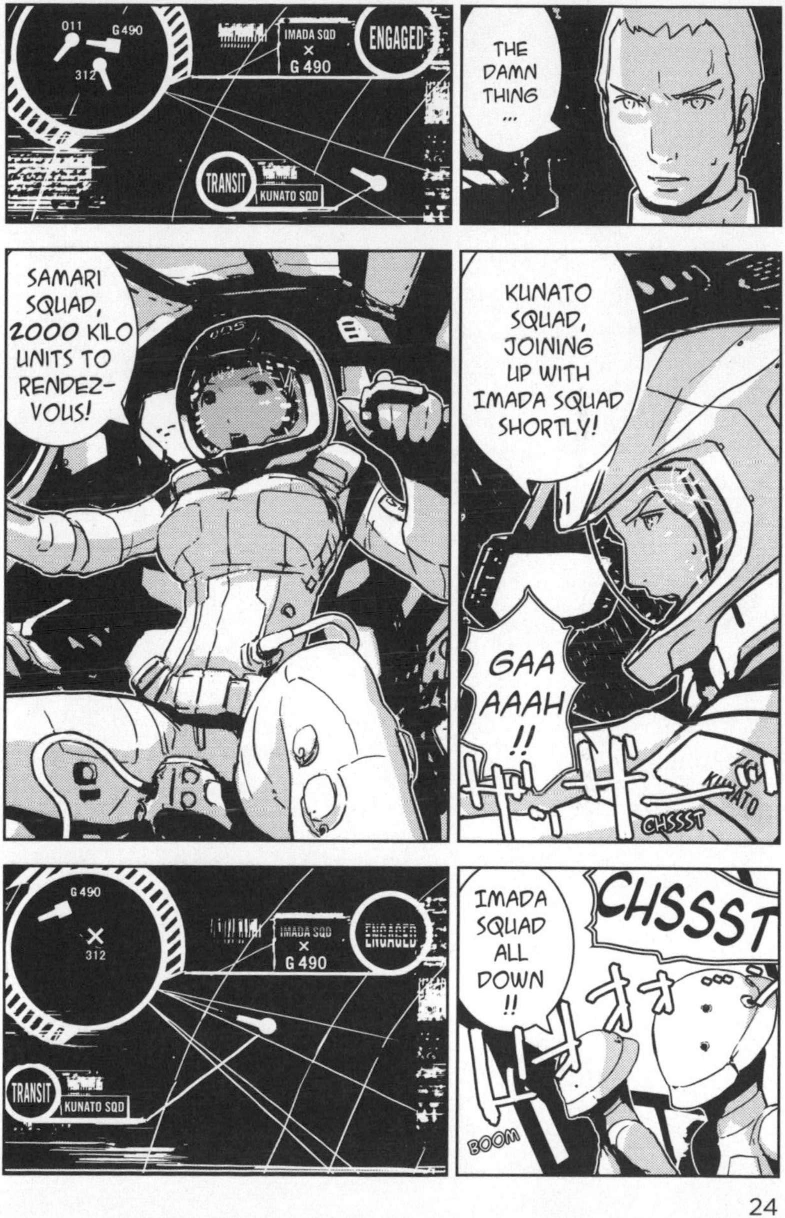
THE DAMN THING ...
011
G490
312
IMADA SQD × G 490
ENGAGED
TRANSIT
KUNATO SQD
KUNATO SQUAD, JOINING UP WITH IMADA SQUAD SHORTLY!
GAA AAAH !!
CHSSST
SAMARI SQUAD, 2000 KILO UNITS TO RENDEZ-VOUS!
IMADA SQUAD ALL DOWN !!
CHSSST
BOOM
G 490
312
IMADA SQD × G 490
TRANSIT
KUNATO SQD

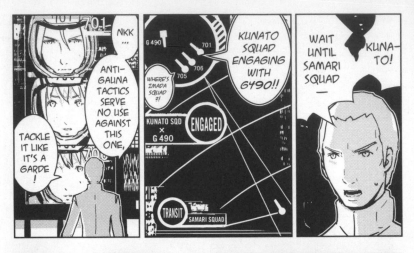

KUNA-TO!
WAIT UNTIL SAMARI SQUAD —
KUNATO SQUAD ENGAGING WITH G490!!
701
G 490
706
705
WHERE'S IMADA SQUAD ?!
KUNATO SQD × G 490
ENGAGED
TRANSIT
SAMARI SQUAD
NKK ...
ANTI-GAUNA TACTICS SERVE NO USE AGAINST THIS ONE,
TACKLE IT LIKE IT'S A GARDE!
ROGER!
701
DON'T DIE ON ME,
YOU GUYS ...

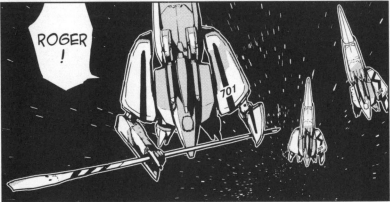

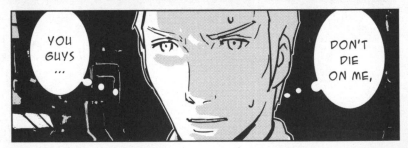

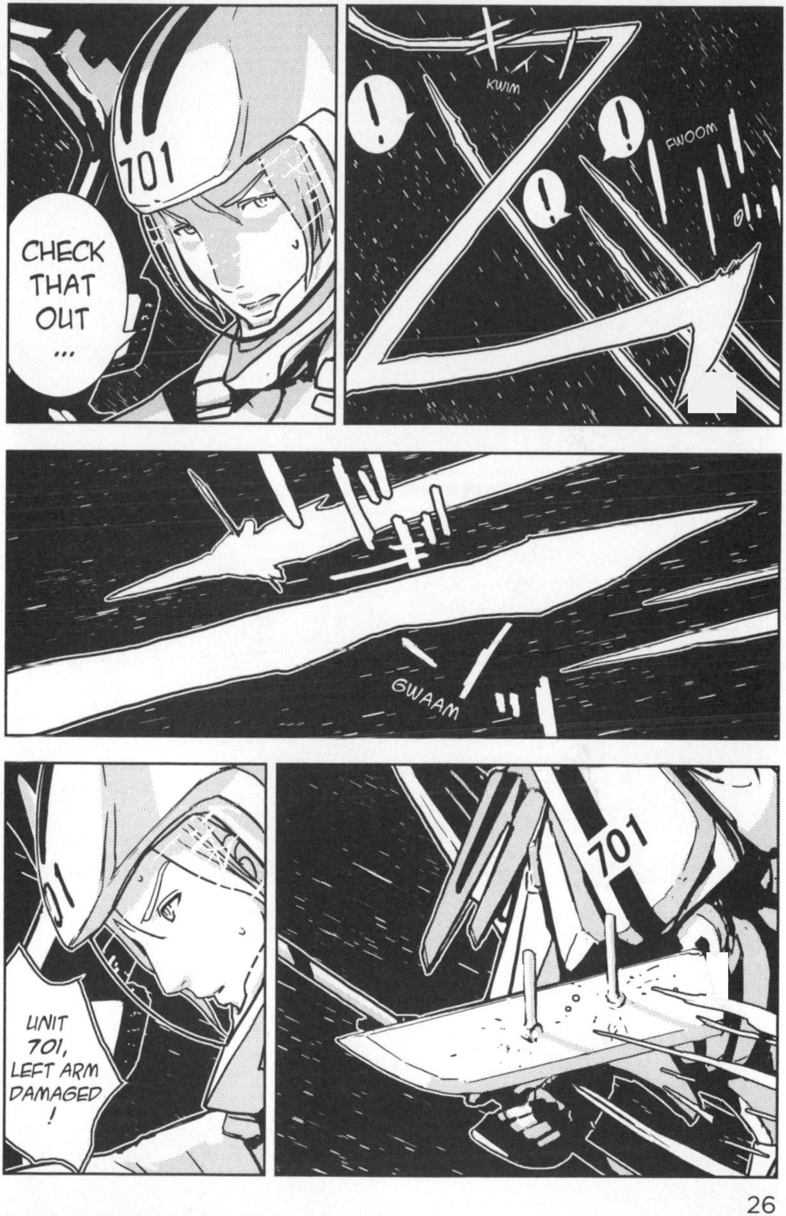
KWIM
!
FWOOM
!
!
CHECK THAT OUT ...
GWAAM
UNIT 701, LEFT ARM DAMAGED!

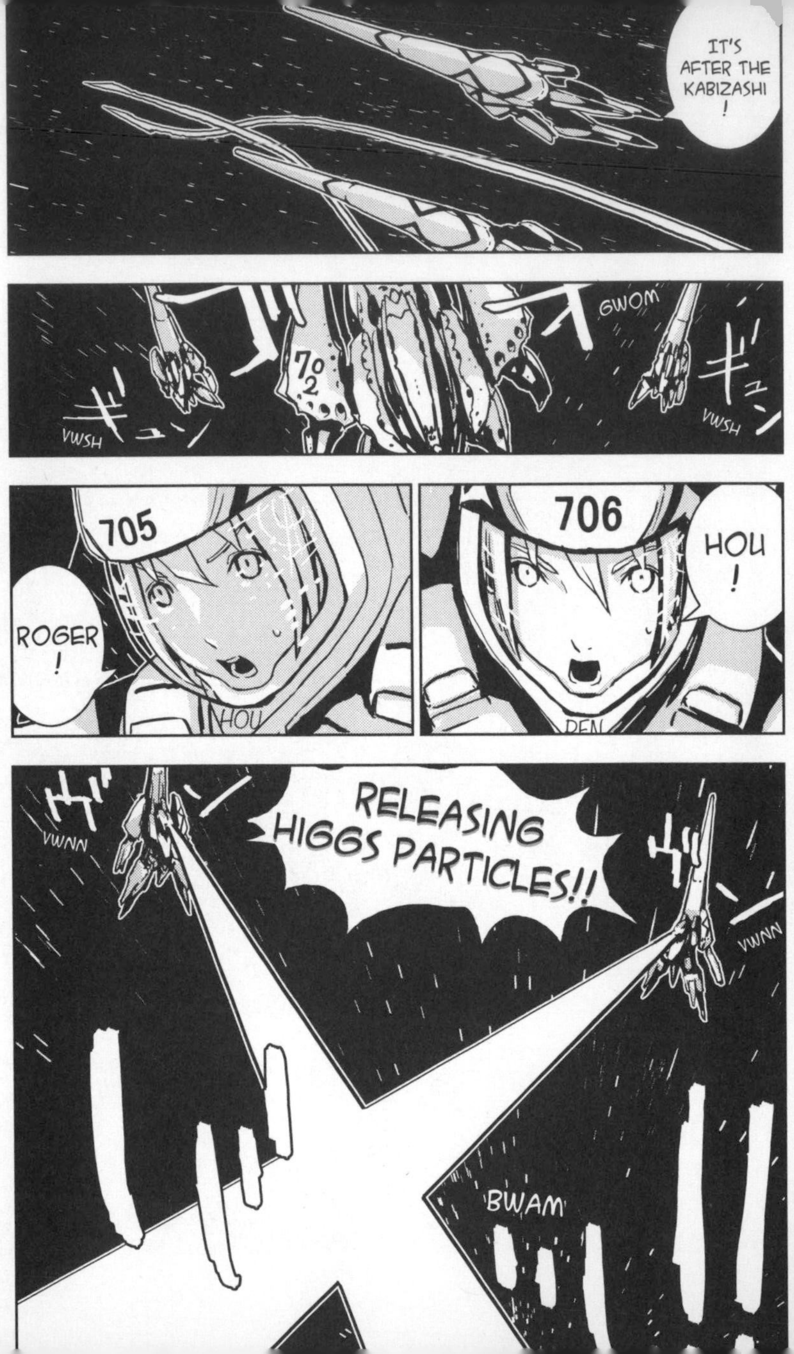
IT'S AFTER THE KABIZASHI!
GWOM
VWSH
702
VWSH
706
HOU!
DEN
705
ROGER!
HOU
RELEASING HIGGS PARTICLES!!
VWNN
VWNN
BWAM

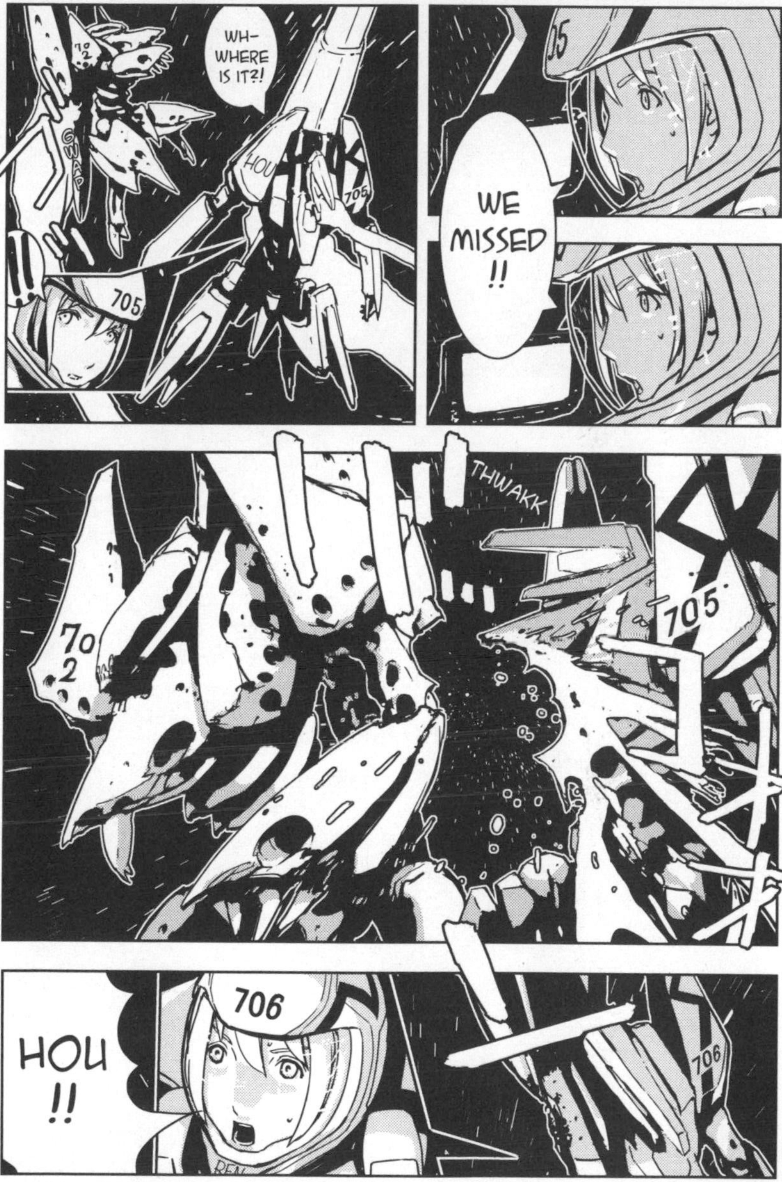
WE MISSED !!
WH- WHERE IS IT?!
GWAP
!!
THWAKK
HOU !!

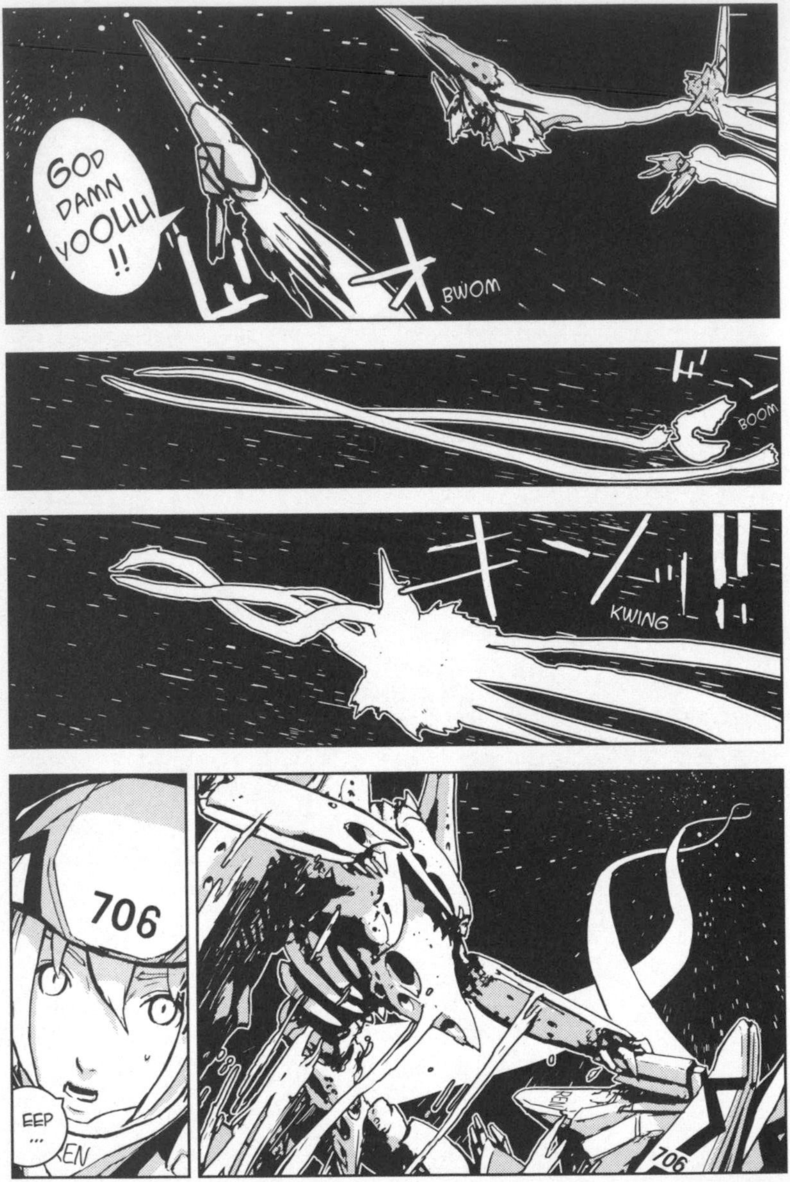
GOD DAMN YOOUU!!
BWOM
BOOM
KWING
706
706
EEP...

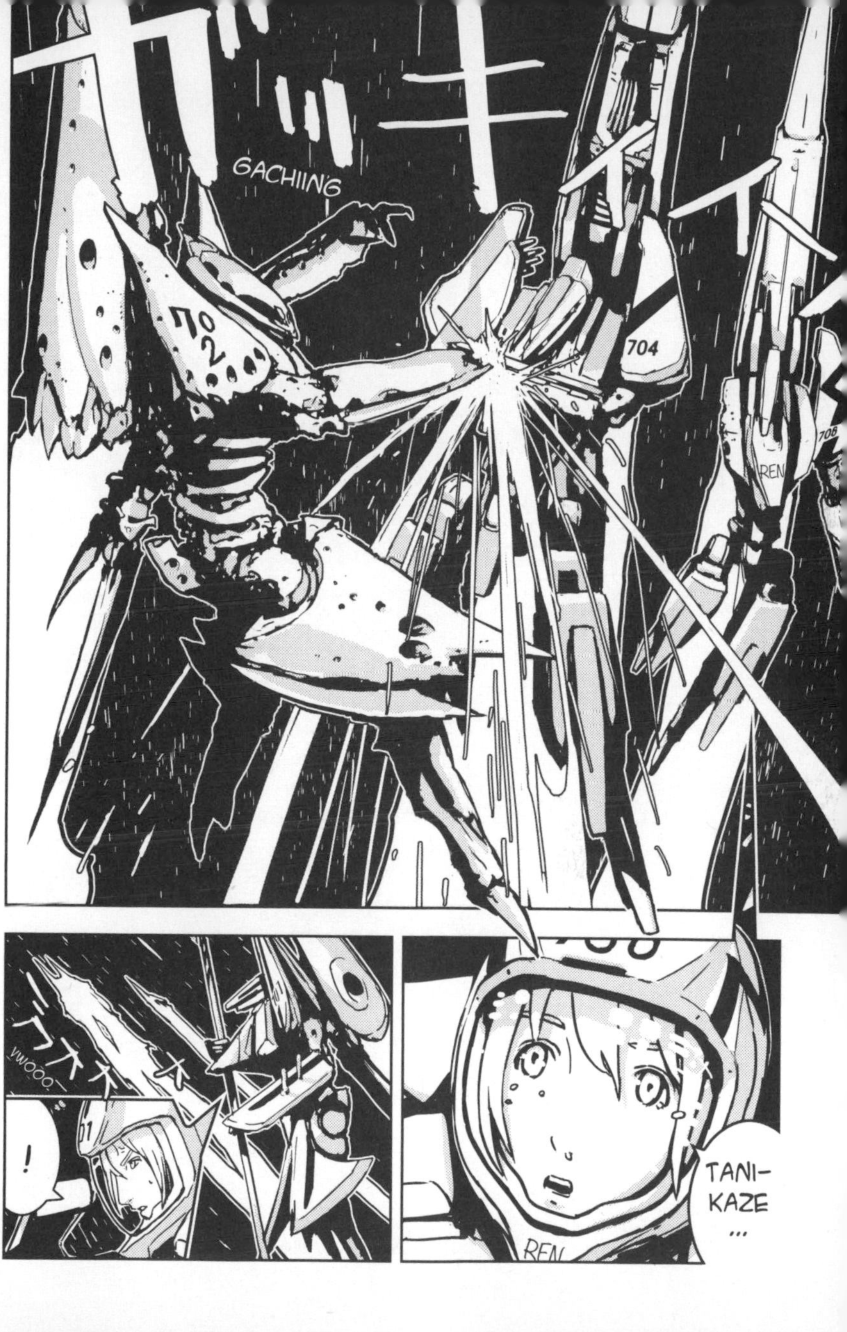
GACHIING
TANI-KAZE ...
YWOOO
!

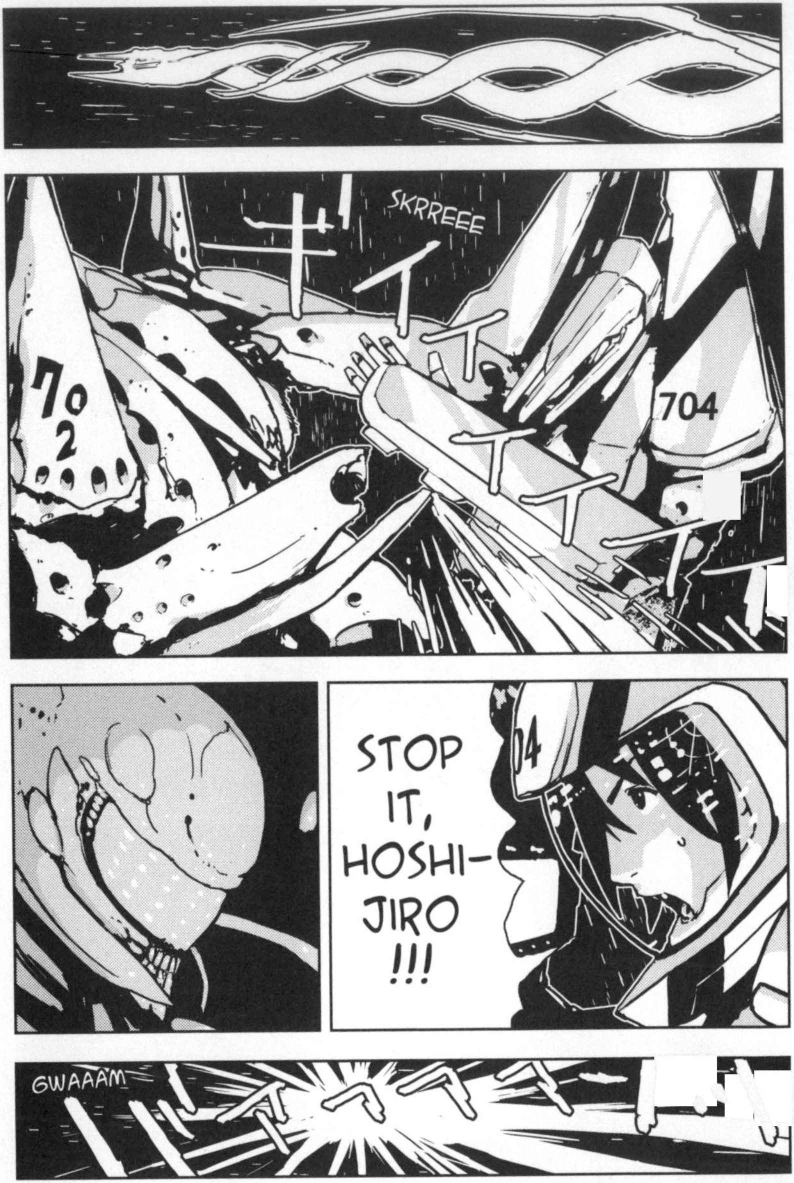
SKRREEE
STOP IT, HOSHI-JIRO !!!
GWAAAM

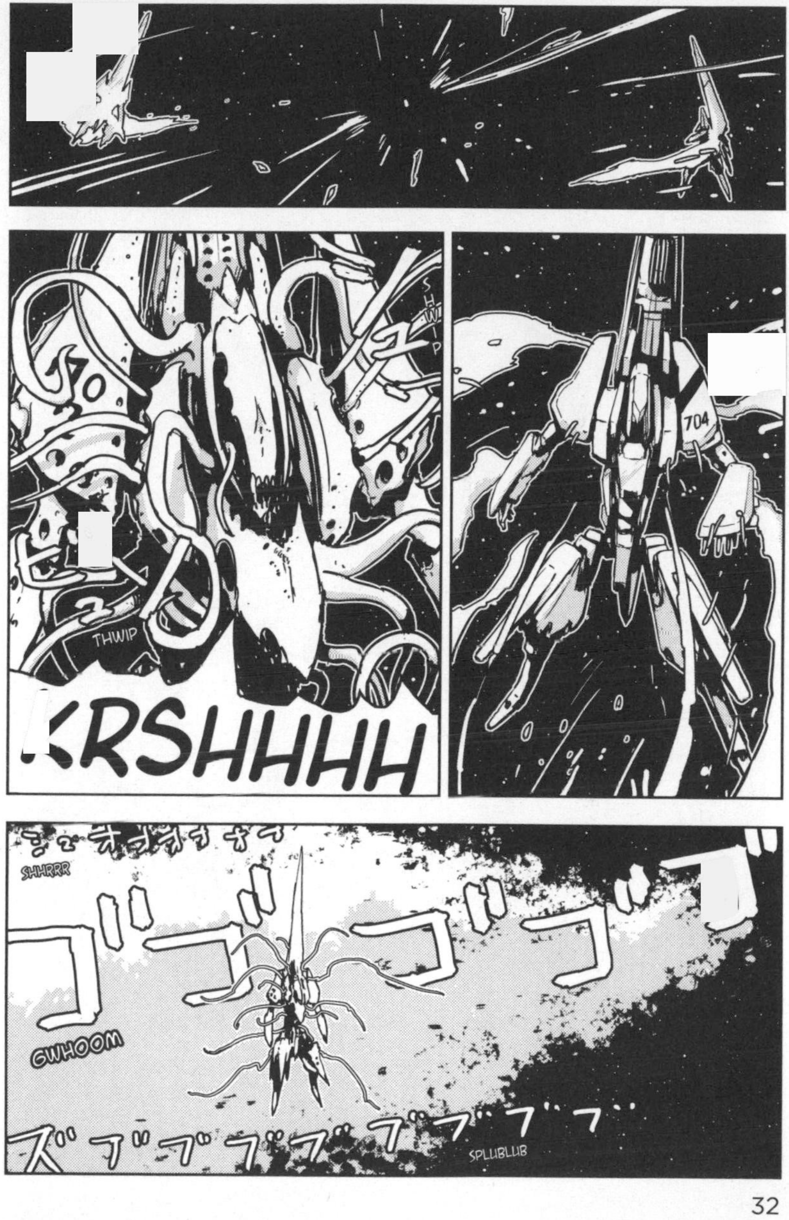
704
SHWIP
THWIP
KRSHHHH
SHHRRR
GWHOOM
SPLUBLUB

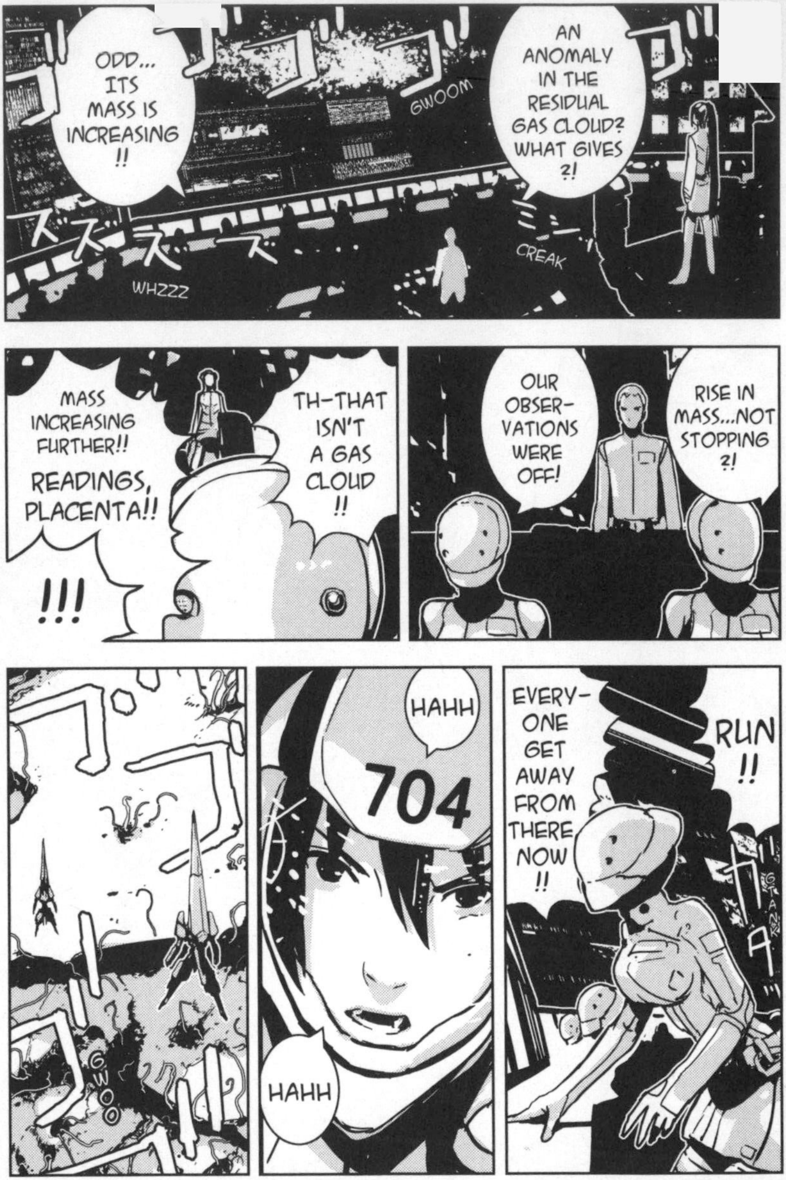
AN ANOMALY IN THE RESIDUAL GAS CLOUD? WHAT GIVES ?!
GWOOM
ODD... ITS MASS IS INCREASING !!
CREAK
WHZZZ
RISE IN MASS...NOT STOPPING ?!
OUR OBSER- VATIONS WERE OFF!
TH-THAT ISN'T A GAS CLOUD !!
MASS INCREASING FURTHER!!
READINGS, PLACENTA!!
!!!
RUN !!
EVERY- ONE GET AWAY FROM THERE NOW !!
GTANK
HAHH
704
HAHH
GWOO

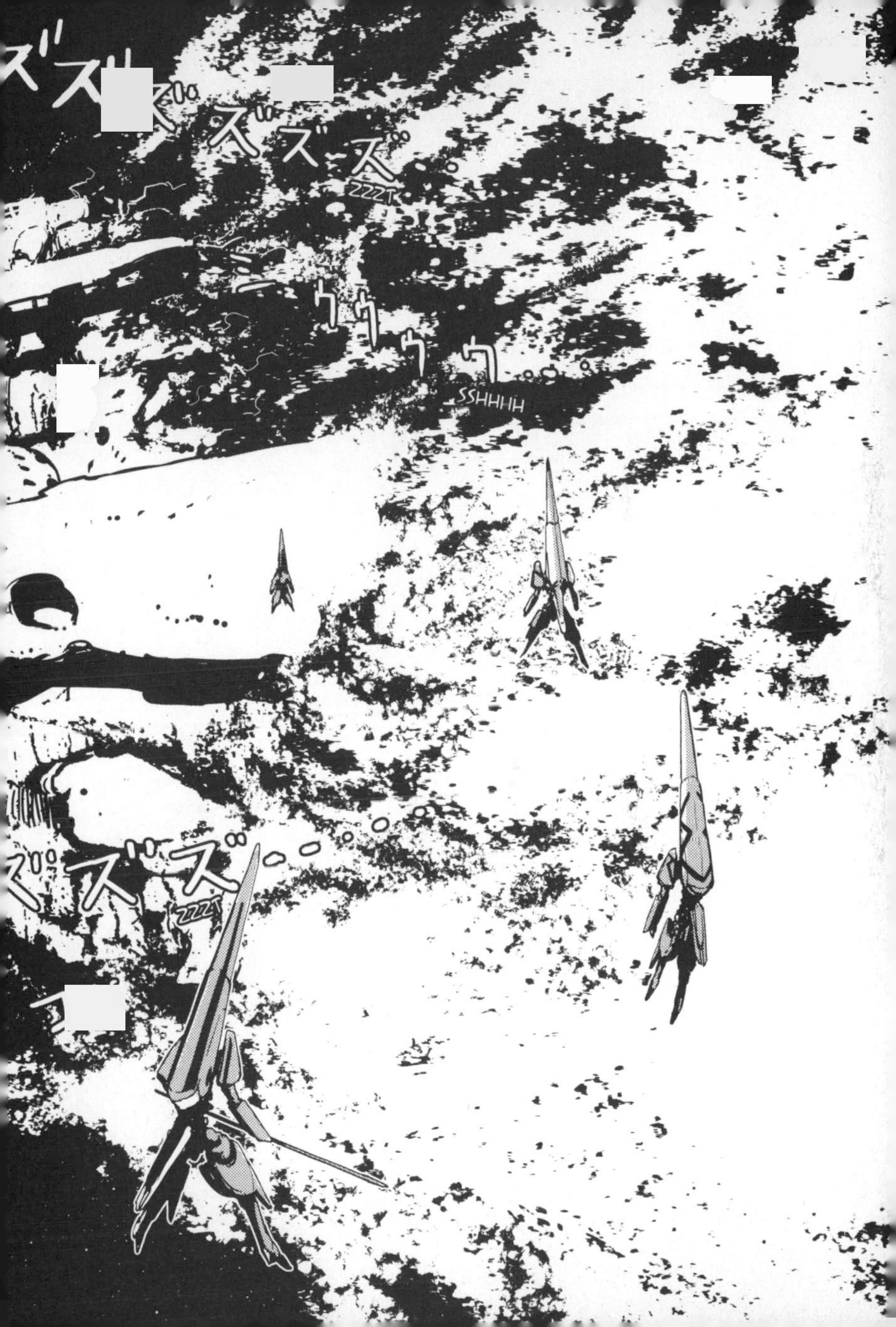
ZZZT
SSHHHH
ZZZT

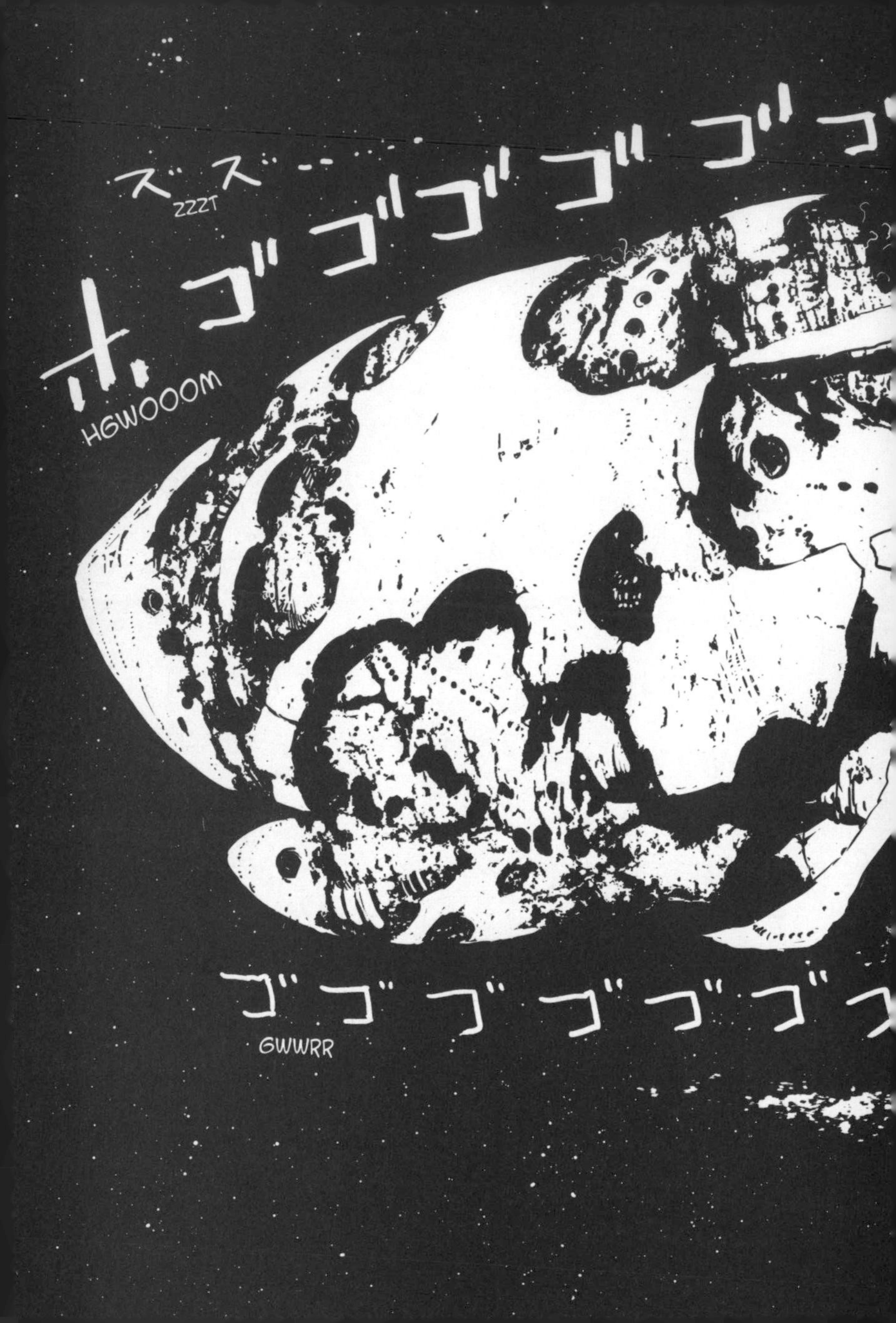
ズズ…
ZZZT
ホゴゴゴゴゴ
HGWOOOM
ゴゴゴゴゴゴ
GWWRR

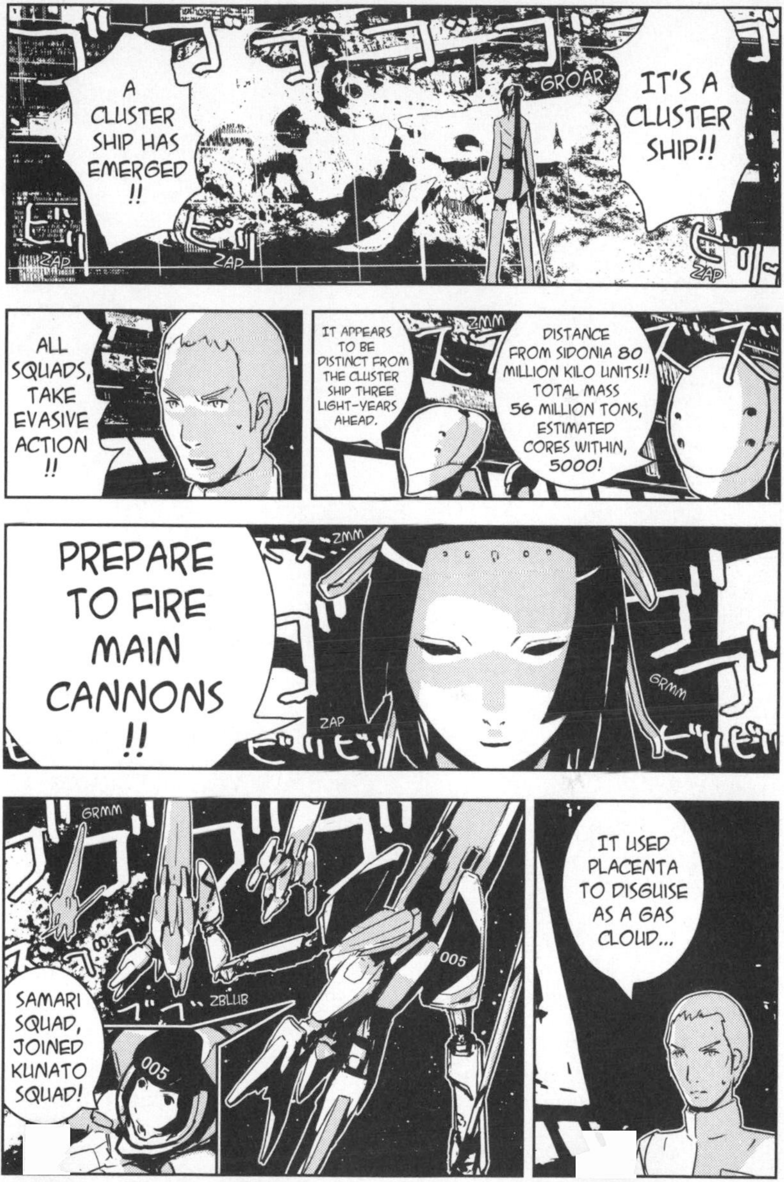

IT'S A CLUSTER SHIP!!
GROAR
A CLUSTER SHIP HAS EMERGED !!
ZAP
ZAP
ZAP
ZMM
DISTANCE FROM SIDONIA 80 MILLION KILO UNITS!! TOTAL MASS 56 MILLION TONS, ESTIMATED CORES WITHIN, 5000!
IT APPEARS TO BE DISTINCT FROM THE CLUSTER SHIP THREE LIGHT-YEARS AHEAD.
ALL SQUADS, TAKE EVASIVE ACTION !!
ZMM
GRMM
ZAP
PREPARE TO FIRE MAIN CANNONS !!
IT USED PLACENTA TO DISGUISE AS A GAS CLOUD...
GRMM
005
ZBLUB
SAMARI SQUAD, JOINED KUNATO SQUAD!
005

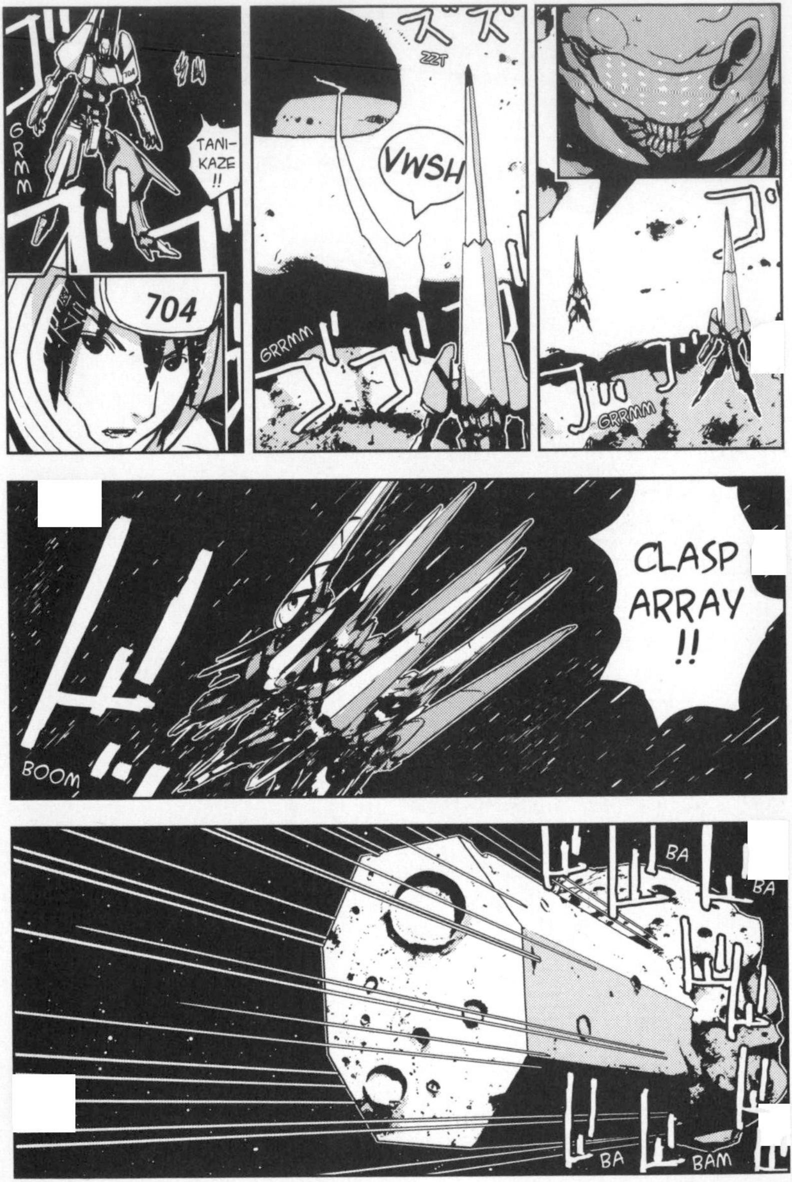
GRRMM
ZZT
VWSH
GRRMM
GRMM
TANI-KAZE !!
704
CLASP ARRAY !!
BOOM
BA
BA
BA
BAM

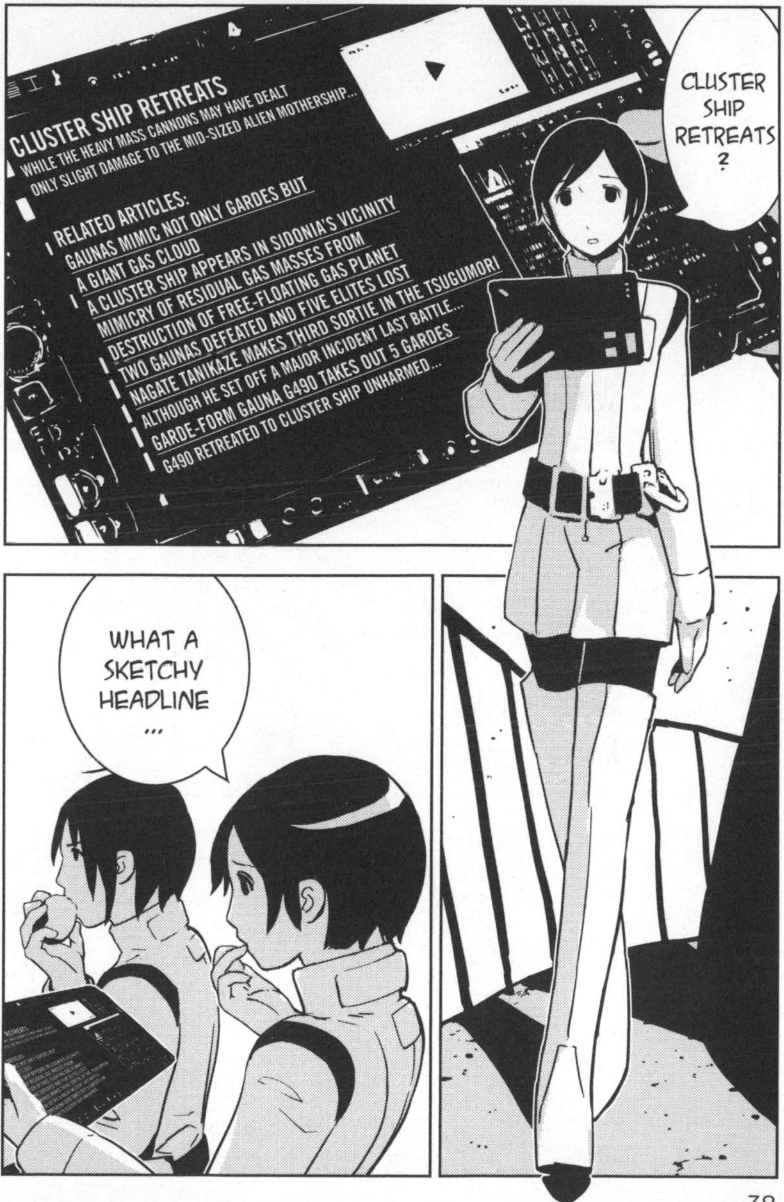
CLUSTER SHIP RETREATS?
CLUSTER SHIP RETREATS
WHILE THE HEAVY MASS CANNONS MAY HAVE DEALT ONLY SLIGHT DAMAGE TO THE MID-SIZED ALIEN MOTHERSHIP...
RELATED ARTICLES:
GAUNAS MIMIC NOT ONLY GARDES BUT
A GIANT GAS CLOUD
A CLUSTER SHIP APPEARS IN SIDONIA'S VICINITY
MIMICRY OF RESIDUAL GAS MASSES FROM
DESTRUCTION OF FREE-FLOATING GAS PLANET
TWO GAUNAS DEFEATED AND FIVE ELITES LOST
NAGATE TANIKAZE MAKES THIRD SORTIE IN THE TSUGUMORI
ALTHOUGH HE SET OFF A MAJOR INCIDENT LAST BATTLE...
GARDE-FORM GAUNA G490 TAKES OUT 5 GARDES
G490 RETREATED TO CLUSTER SHIP UNHARMED...
WHAT A SKETCHY HEADLINE ...

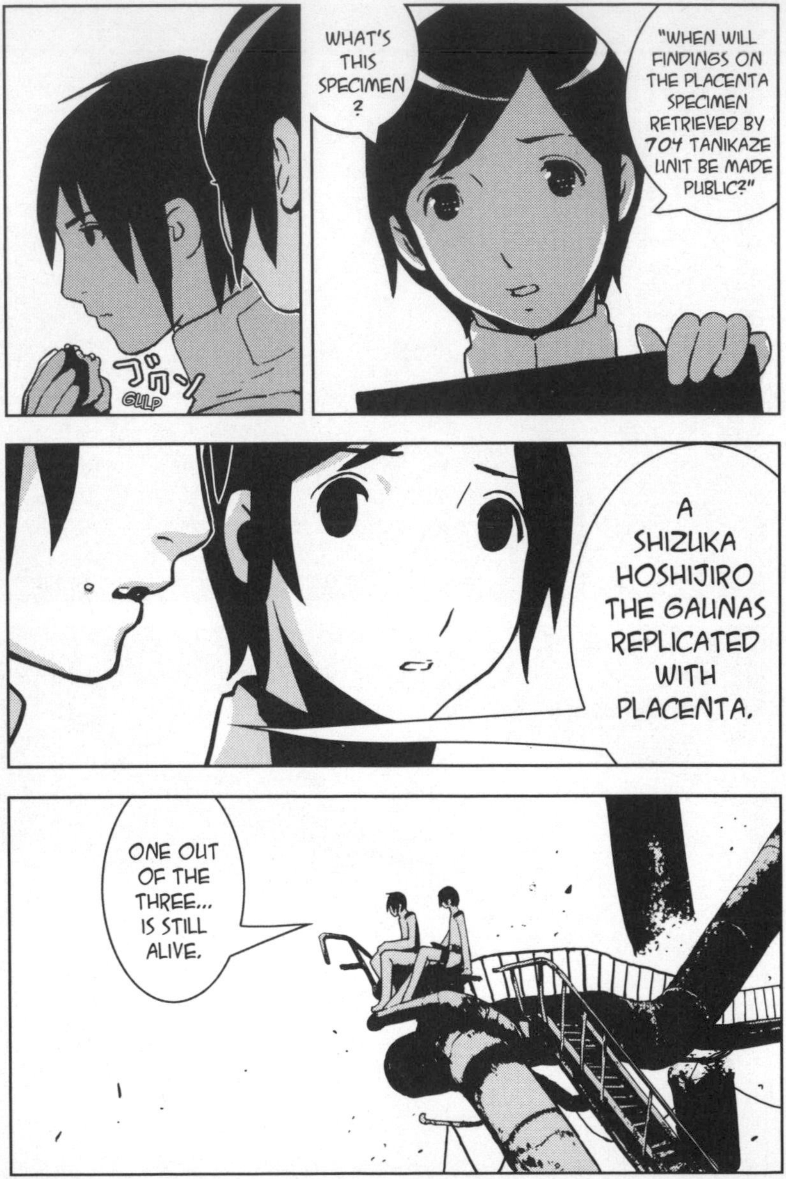
"WHEN WILL FINDINGS ON THE PLACENTA SPECIMEN RETRIEVED BY 704 TANIKAZE UNIT BE MADE PUBLIC?"
WHAT'S THIS SPECIMEN?
GULP
A SHIZUKA HOSHIJIRO THE GAUNAS REPLICATED WITH PLACENTA.
ONE OUT OF THE THREE... IS STILL ALIVE.

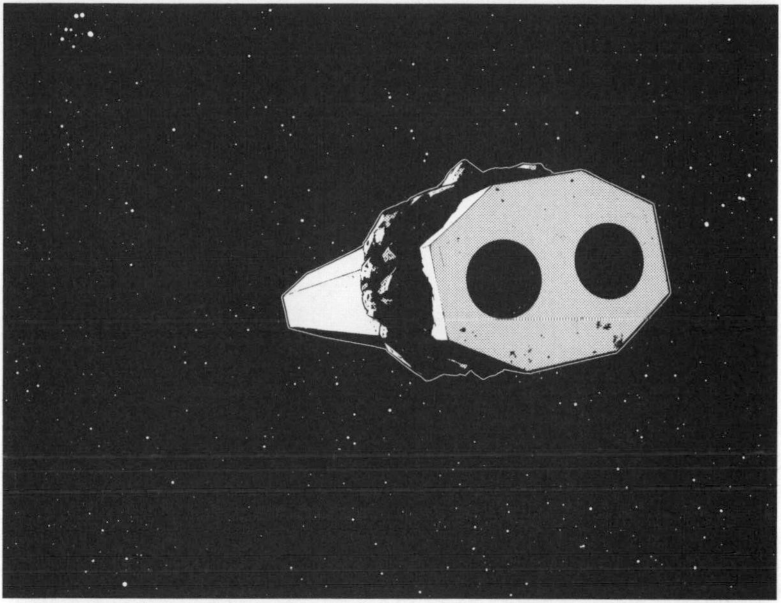

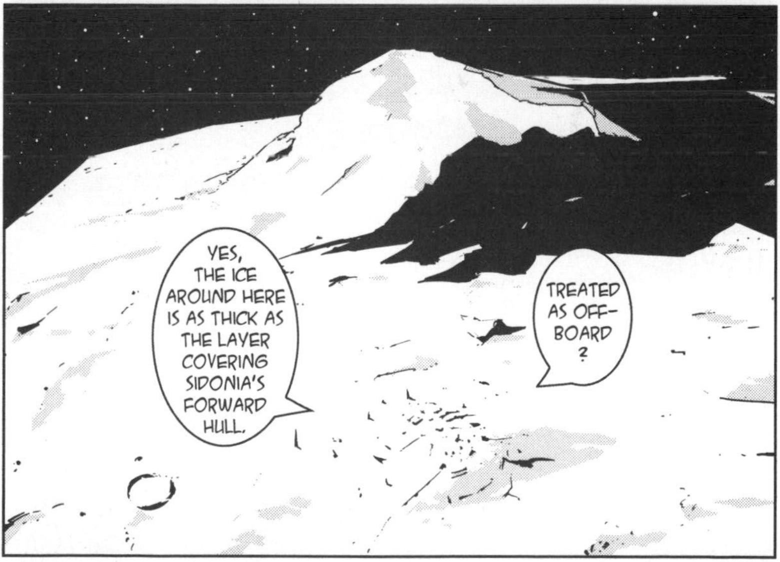
TREATED AS OFF-BOARD?
YES, THE ICE AROUND HERE IS AS THICK AS THE LAYER COVERING SIDONIA'S FORWARD HULL.

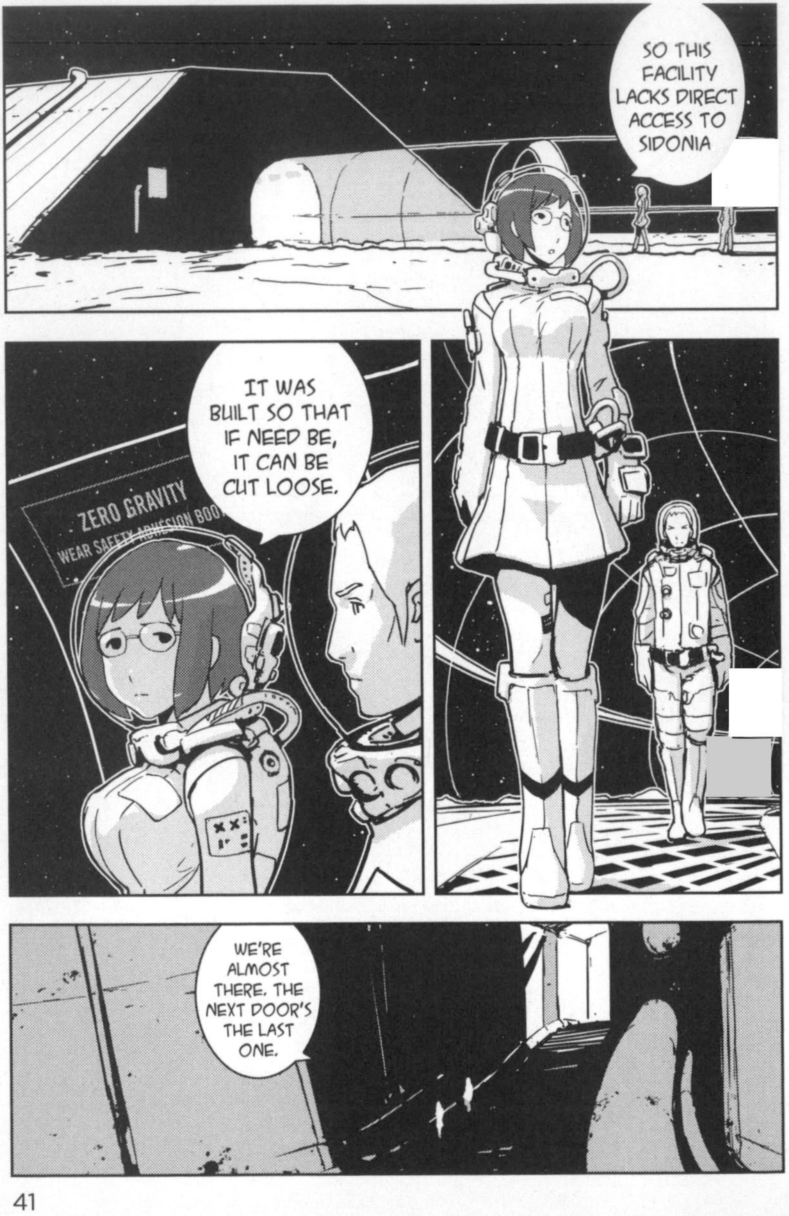
SO THIS FACILITY LACKS DIRECT ACCESS TO SIDONIA
IT WAS BUILT SO THAT IF NEED BE, IT CAN BE CUT LOOSE.
ZERO GRAVITY
WE'RE ALMOST THERE. THE NEXT DOOR'S THE LAST ONE.

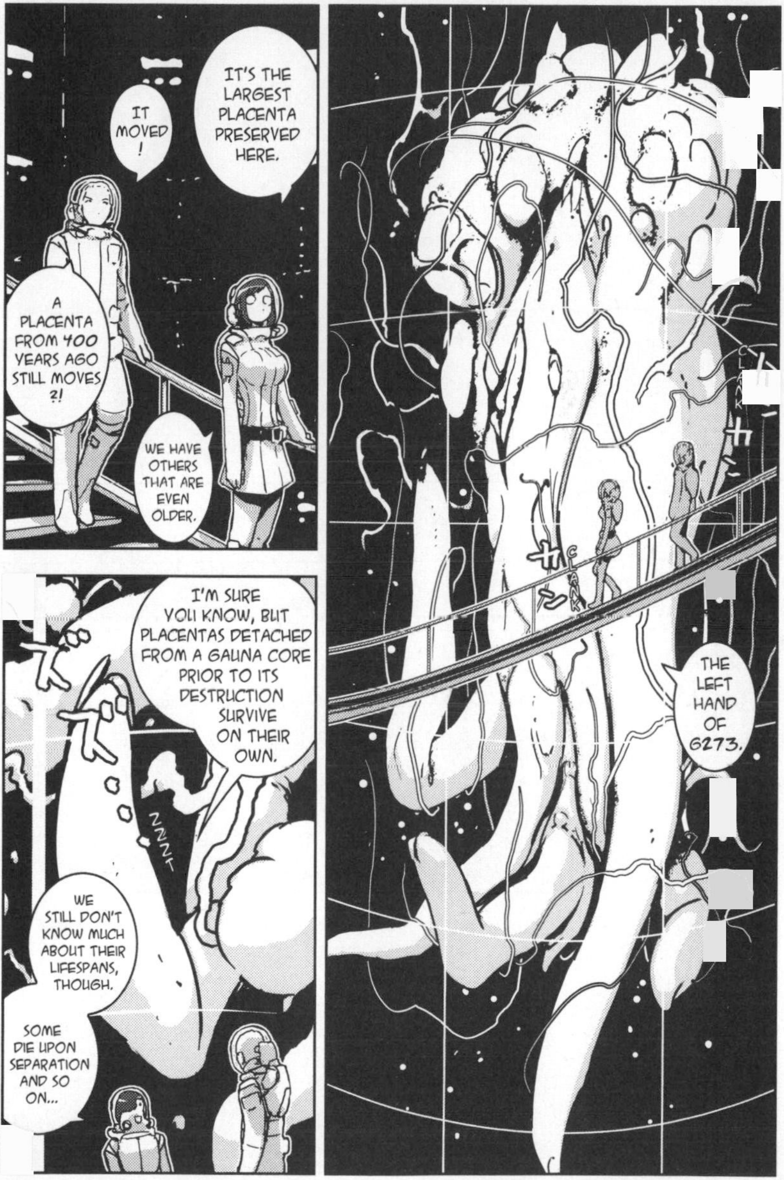
CLANK
CLANK
THE LEFT HAND OF 6273.
IT'S THE LARGEST PLACENTA PRESERVED HERE.
IT MOVED!
A PLACENTA FROM 400 YEARS AGO STILL MOVES?!
WE HAVE OTHERS THAT ARE EVEN OLDER.
I'M SURE YOU KNOW, BUT PLACENTAS DETACHED FROM A GAUNA CORE PRIOR TO ITS DESTRUCTION SURVIVE ON THEIR OWN.
ZZZT
WE STILL DON'T KNOW MUCH ABOUT THEIR LIFESPANS, THOUGH.
SOME DIE UPON SEPARATION AND SO ON...

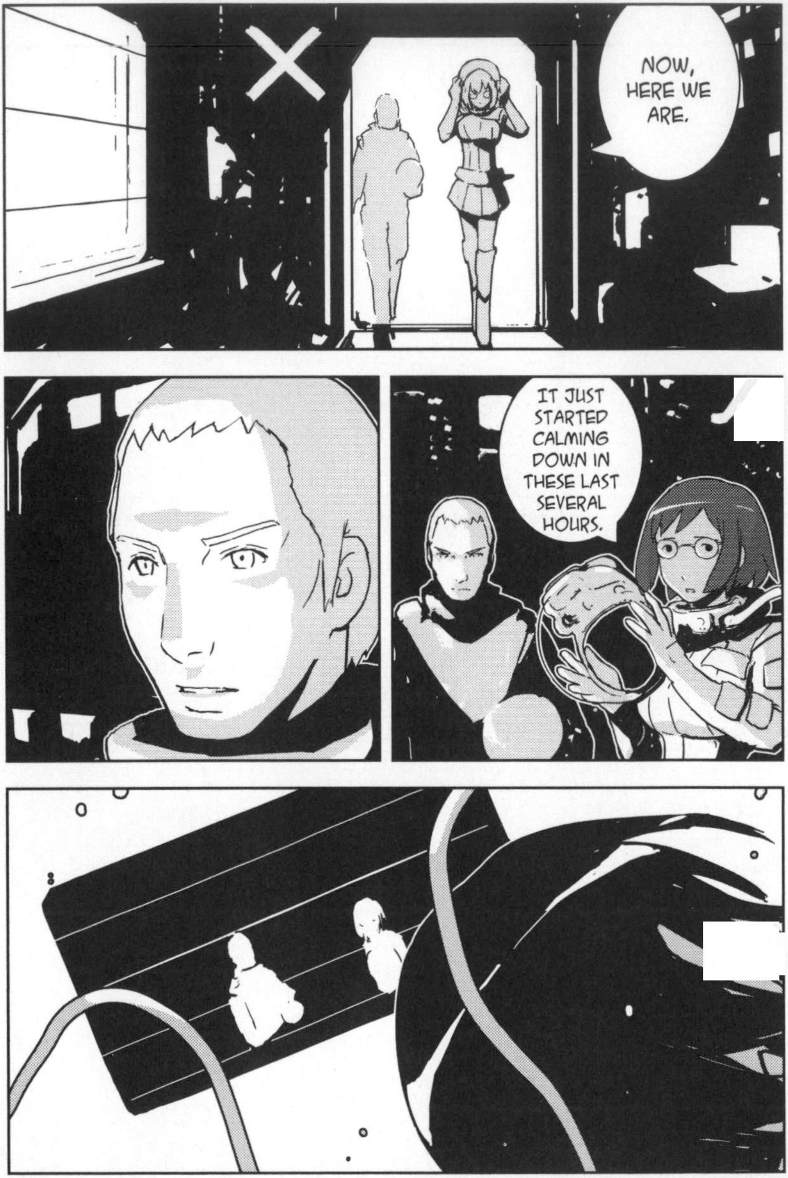
NOW, HERE WE ARE.
IT JUST STARTED CALMING DOWN IN THESE LAST SEVERAL HOURS.

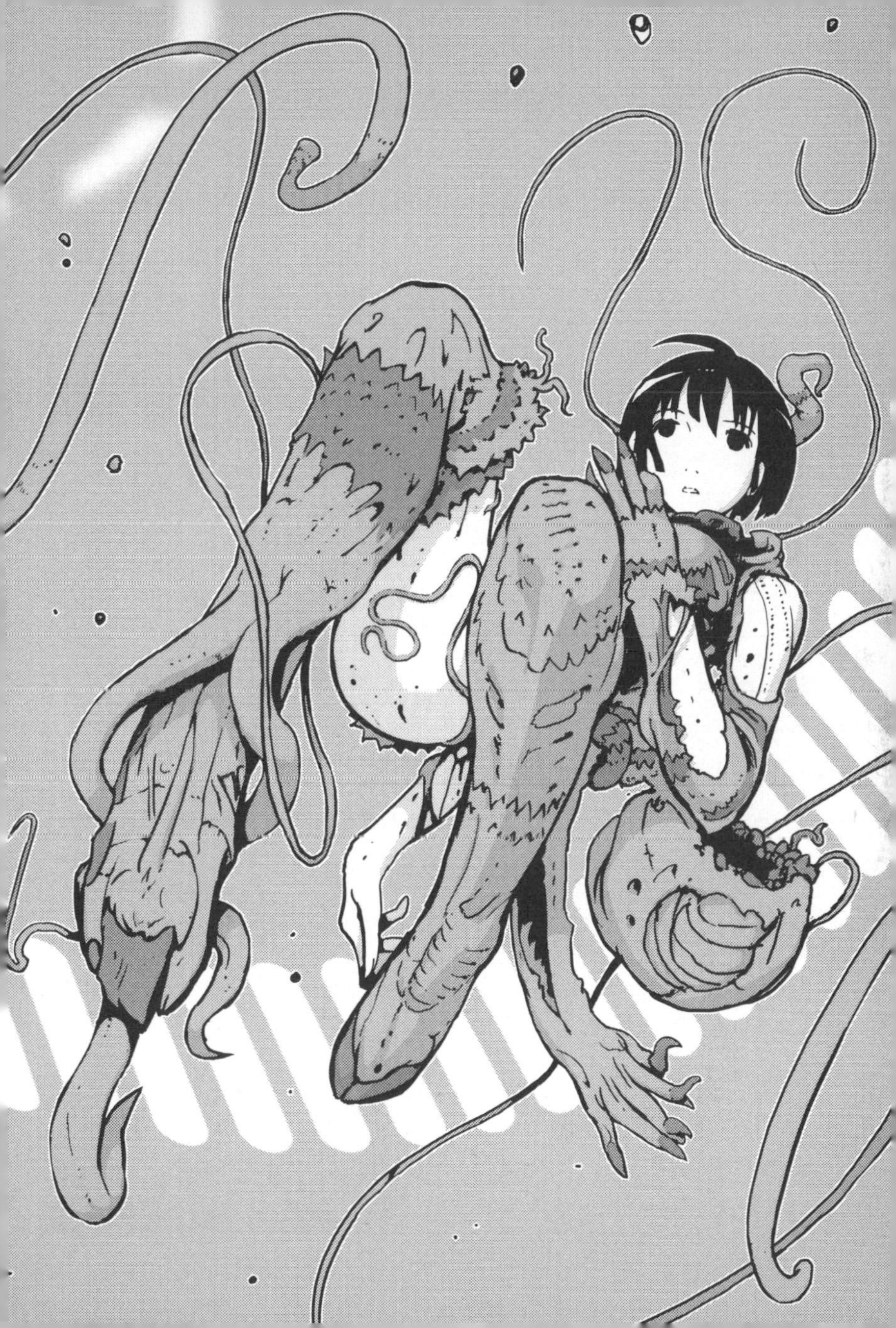

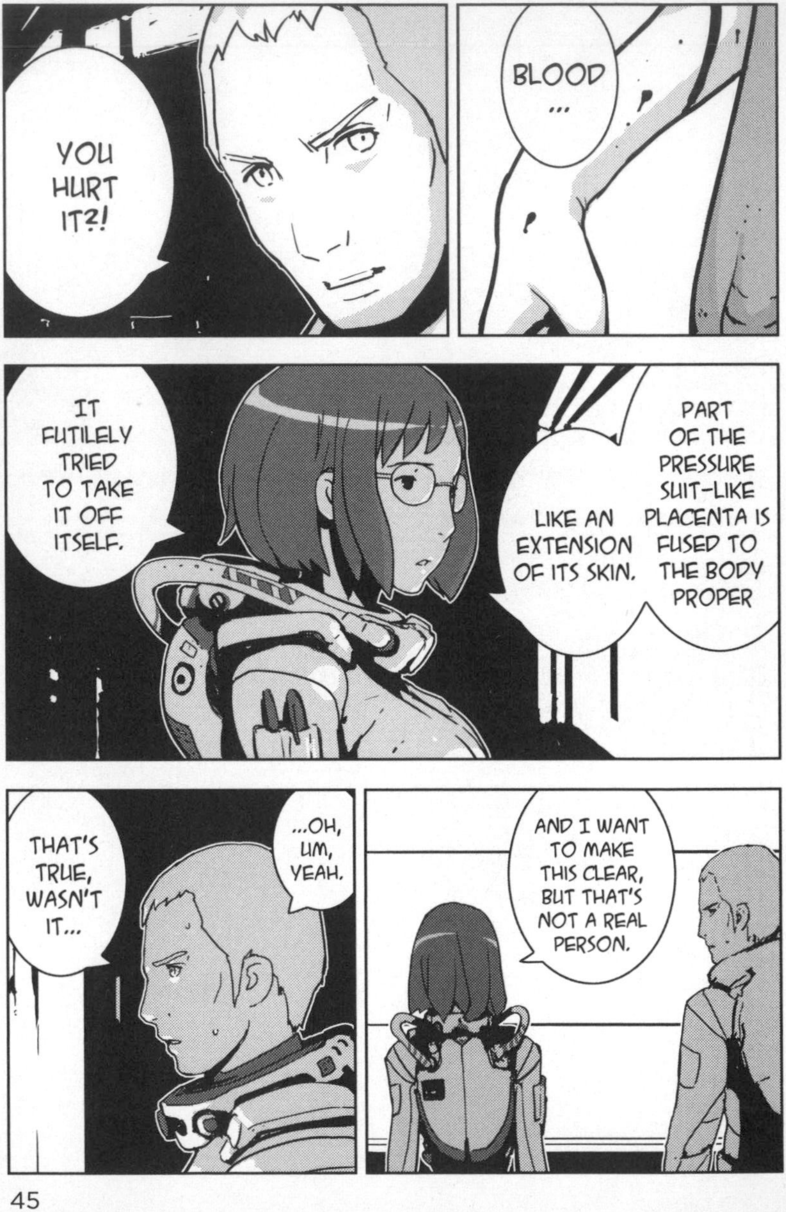
BLOOD ...
YOU HURT IT?!
PART OF THE PRESSURE SUIT-LIKE PLACENTA IS FUSED TO THE BODY PROPER
LIKE AN EXTENSION OF ITS SKIN.
IT FUTILELY TRIED TO TAKE IT OFF ITSELF.
AND I WANT TO MAKE THIS CLEAR, BUT THAT'S NOT A REAL PERSON.
...OH, UM, YEAH.
THAT'S TRUE, WASN'T IT...

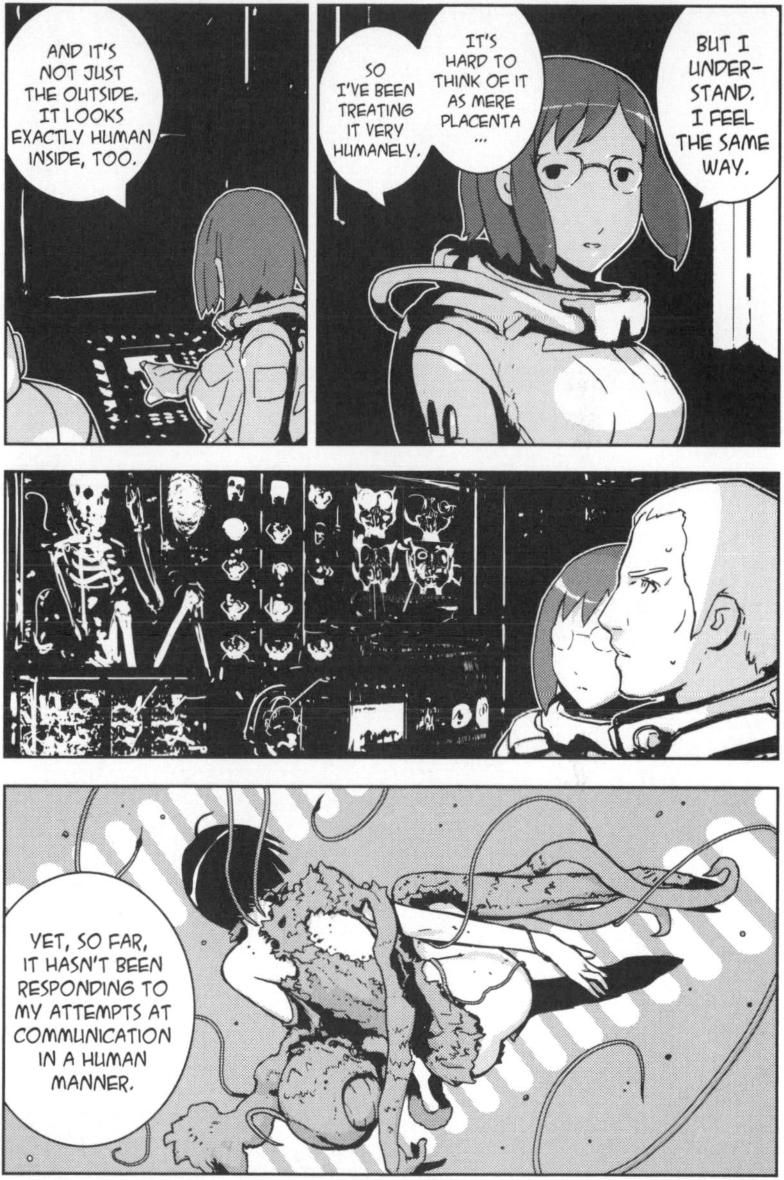
BUT I UNDER-STAND. I FEEL THE SAME WAY.
IT'S HARD TO THINK OF IT AS MERE PLACENTA ...
SO I'VE BEEN TREATING IT VERY HUMANELY.
AND IT'S NOT JUST THE OUTSIDE. IT LOOKS EXACTLY HUMAN INSIDE, TOO.
YET, SO FAR, IT HASN'T BEEN RESPONDING TO MY ATTEMPTS AT COMMUNICATION IN A HUMAN MANNER.

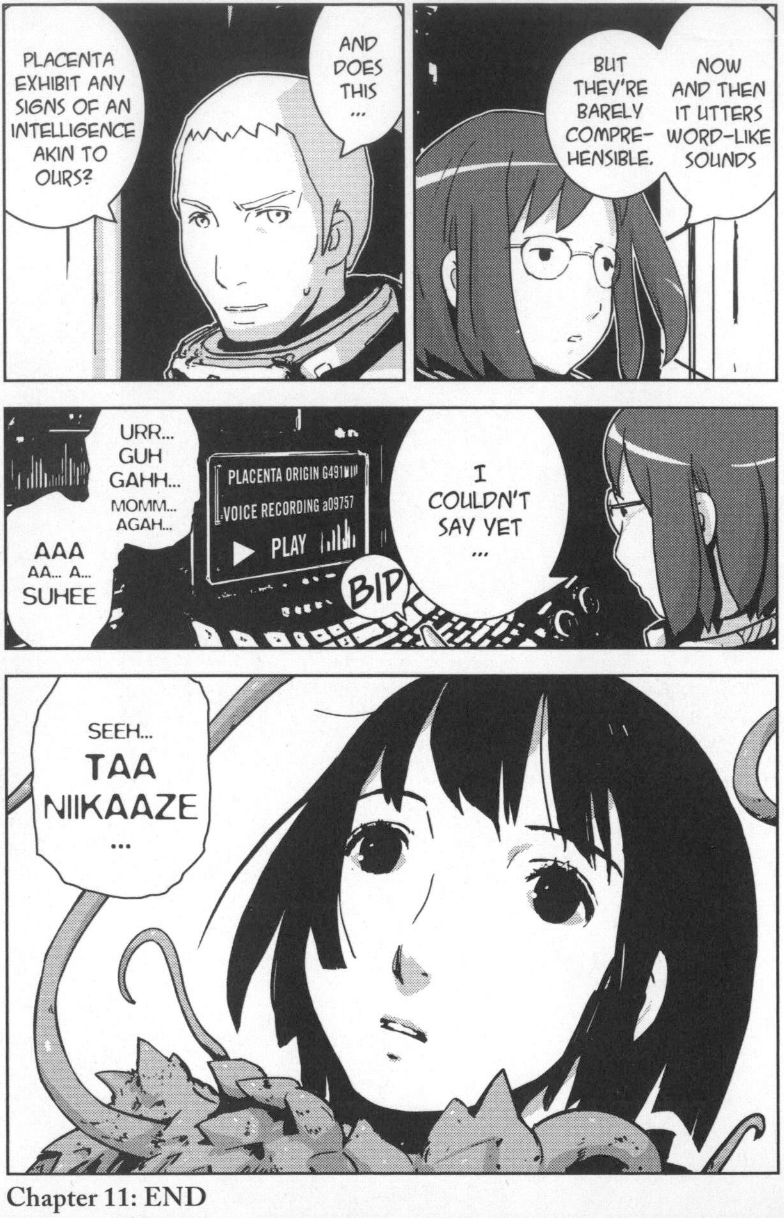
NOW AND THEN IT UTTERS WORD-LIKE SOUNDS
BUT THEY'RE BARELY COMPRE-HENSIBLE.
AND DOES THIS ...
PLACENTA EXHIBIT ANY SIGNS OF AN INTELLIGENCE AKIN TO OURS?
I COULDN'T SAY YET ...
BIP
PLACENTA ORIGIN G491
VOICE RECORDING a09757
PLAY
URR... GUH GAHH... MOMM... AGAH...
AAA AA... A... SUHEE
SEEH... TAA NIIKAAZE ...

Chapter 11: END

Honoka Series

Clones rapidly matured in artificial amniotic fluid.

Age: five.

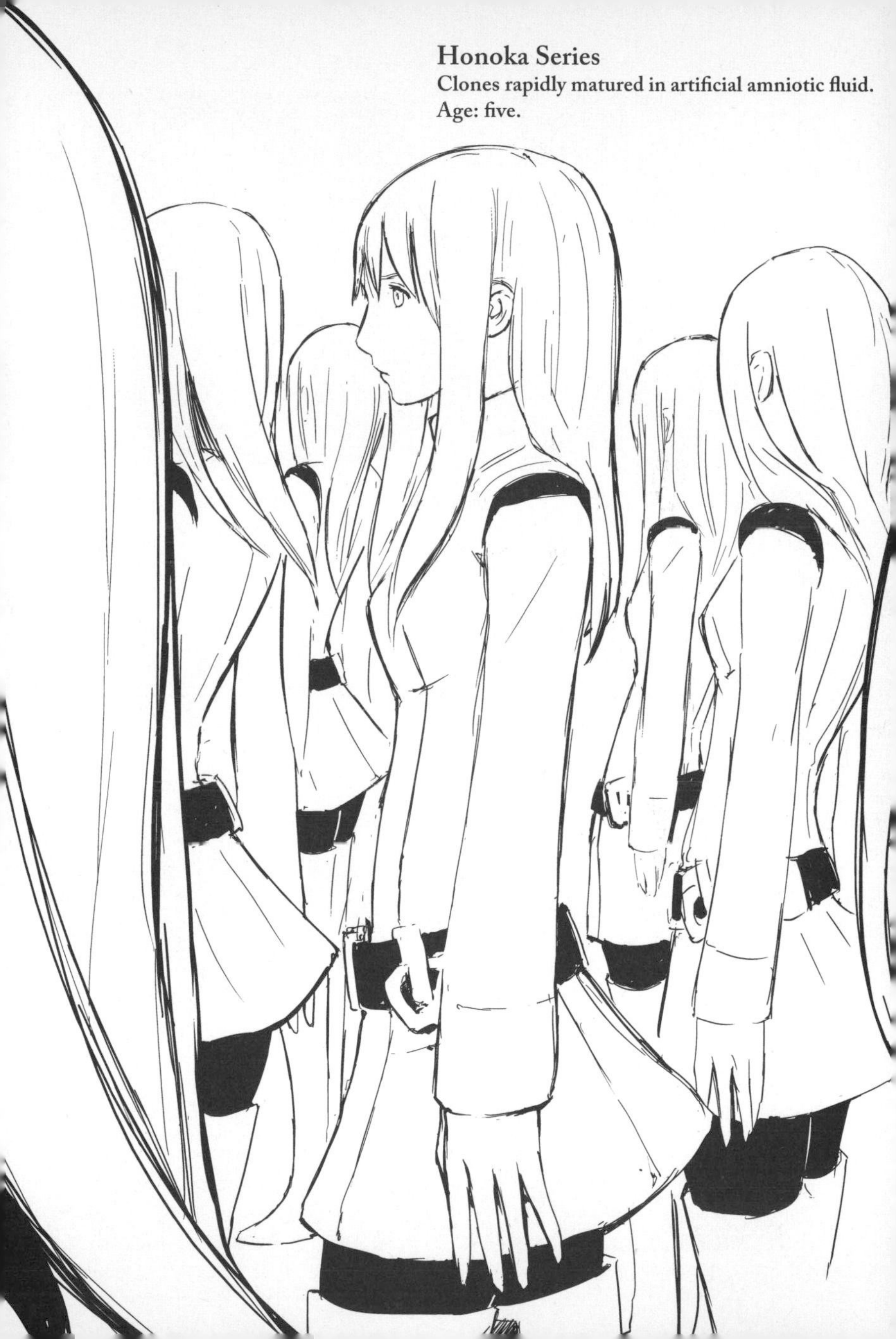

Chapter 12: Numi Tahiro's Weekly Meal

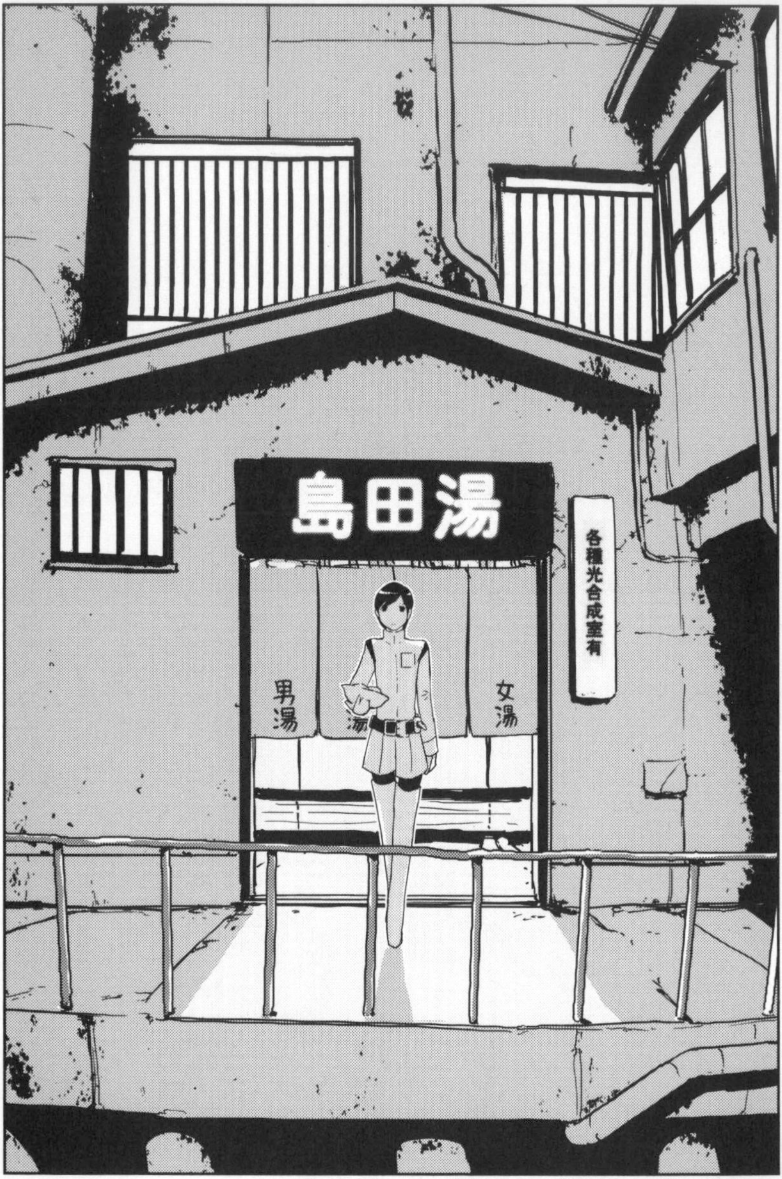
島田湯
各種光合成室有
男湯
湯
女湯

One Hundred Sights of Sidonia Part Nine:
Shimada Public Bath House

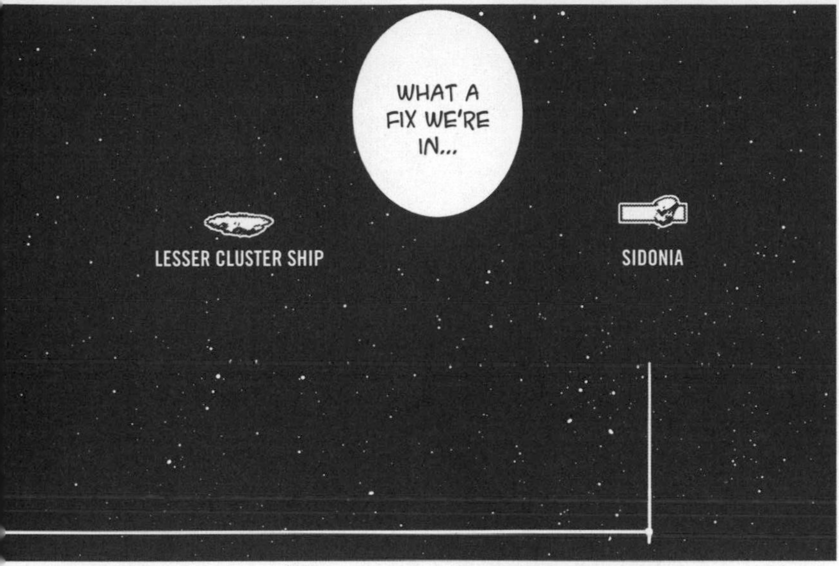
WHAT A FIX WE'RE IN...
LESSER CLUSTER SHIP
SIDONIA

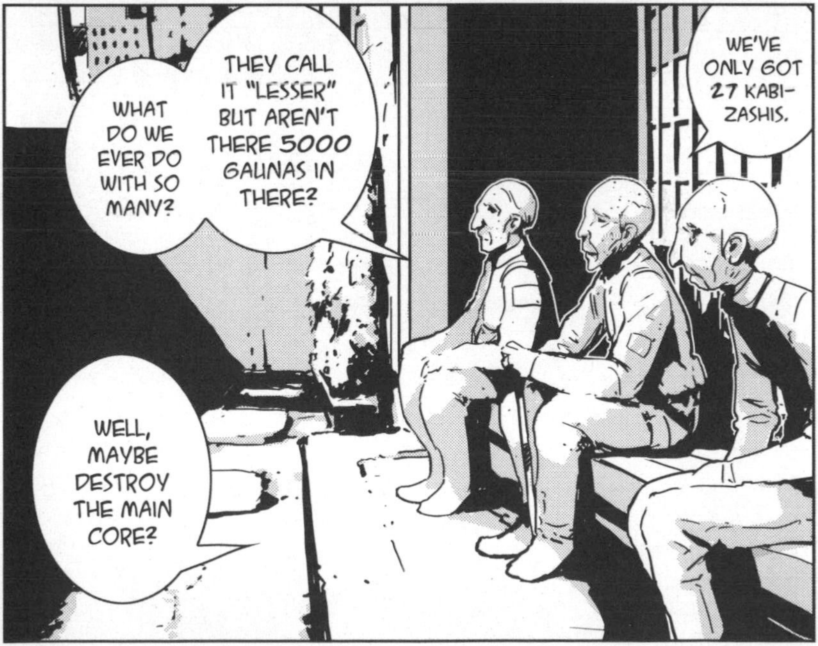
WE'VE ONLY GOT 27 KABI-ZASHIS.
THEY CALL IT "LESSER" BUT AREN'T THERE 5000 GAUNAS IN THERE?
WHAT DO WE EVER DO WITH SO MANY?
WELL, MAYBE DESTROY THE MAIN CORE?

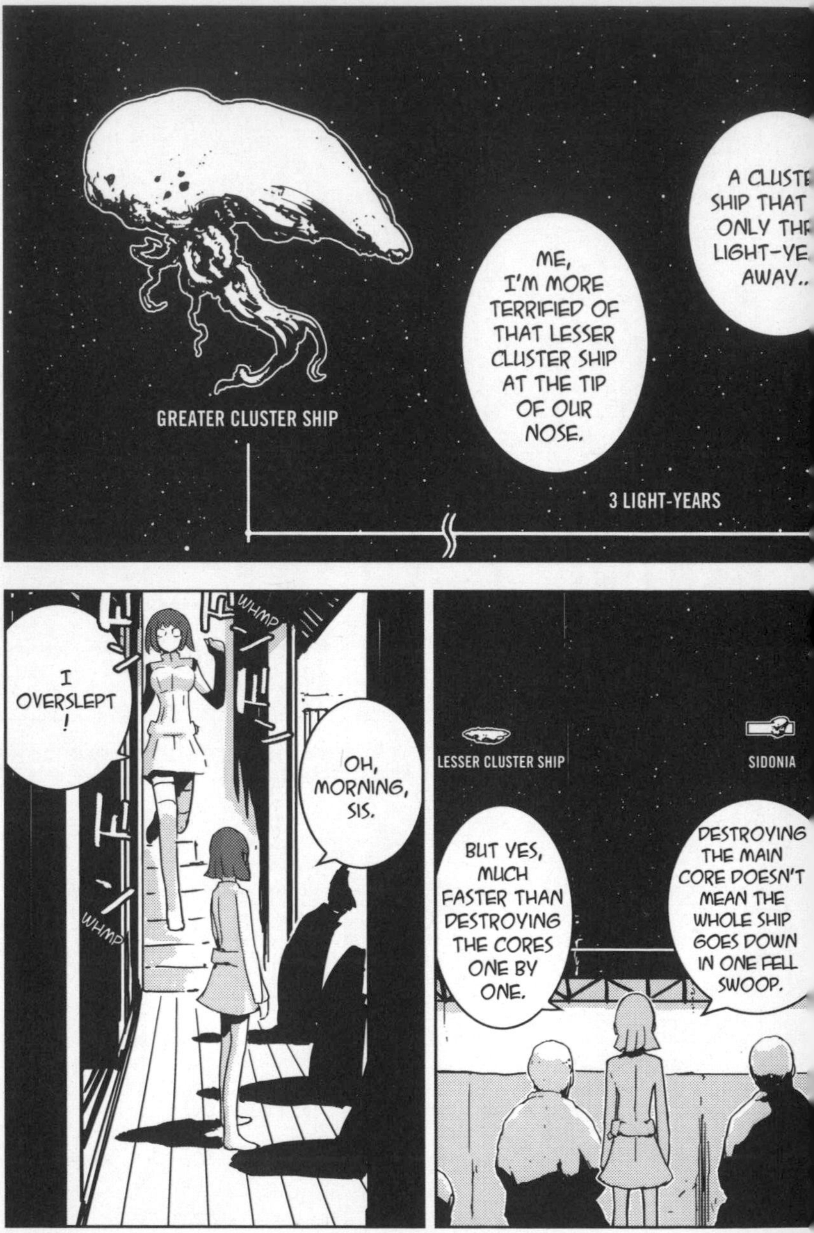

A CLUSTE SHIP THAT ONLY TH LIGHT-YE AWAY..
ME, I'M MORE TERRIFIED OF THAT LESSER CLUSTER SHIP AT THE TIP OF OUR NOSE.
GREATER CLUSTER SHIP
3 LIGHT-YEARS
SIDONIA
LESSER CLUSTER SHIP
DESTROYING THE MAIN CORE DOESN'T MEAN THE WHOLE SHIP GOES DOWN IN ONE FELL SWOOP.
BUT YES, MUCH FASTER THAN DESTROYING THE CORES ONE BY ONE.
WHMP
I OVERSLEPT!
OH, MORNING, SIS.
WHMP

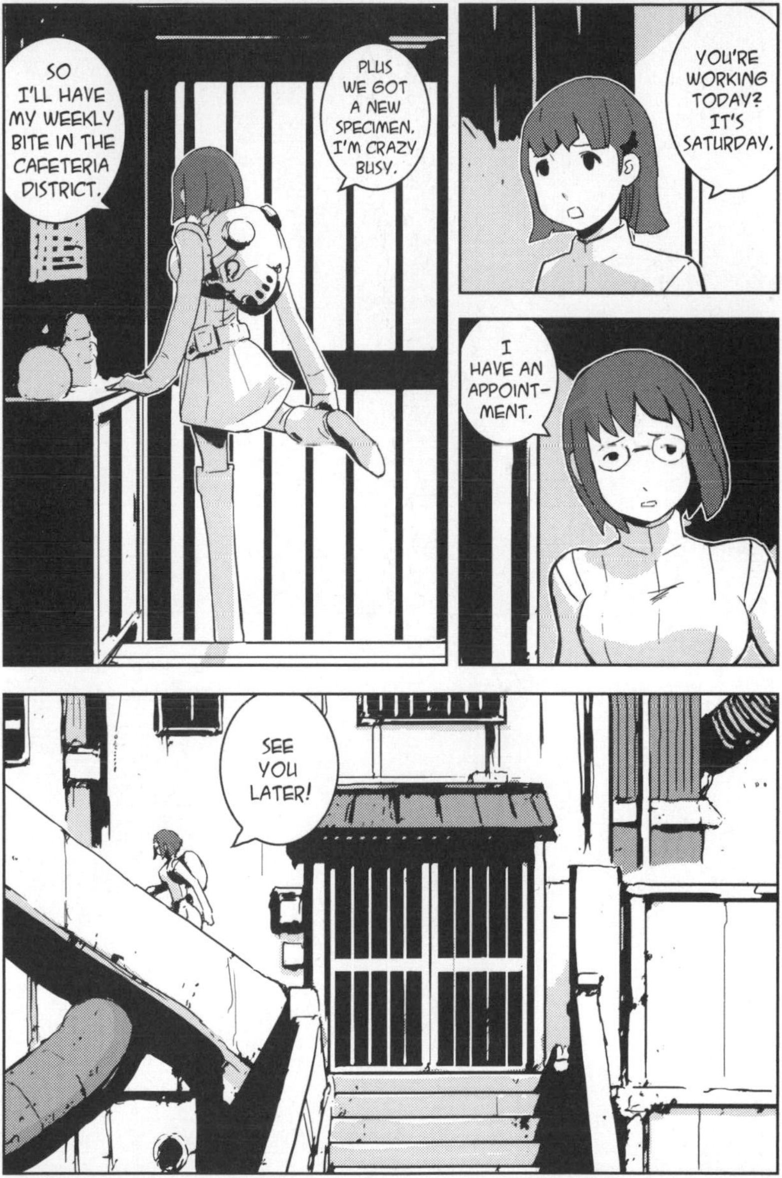
YOU'RE WORKING TODAY? IT'S SATURDAY.
I HAVE AN APPOINT-MENT.
PLUS WE GOT A NEW SPECIMEN. I'M CRAZY BUSY.
SO I'LL HAVE MY WEEKLY BITE IN THE CAFETERIA DISTRICT.
SEE YOU LATER!

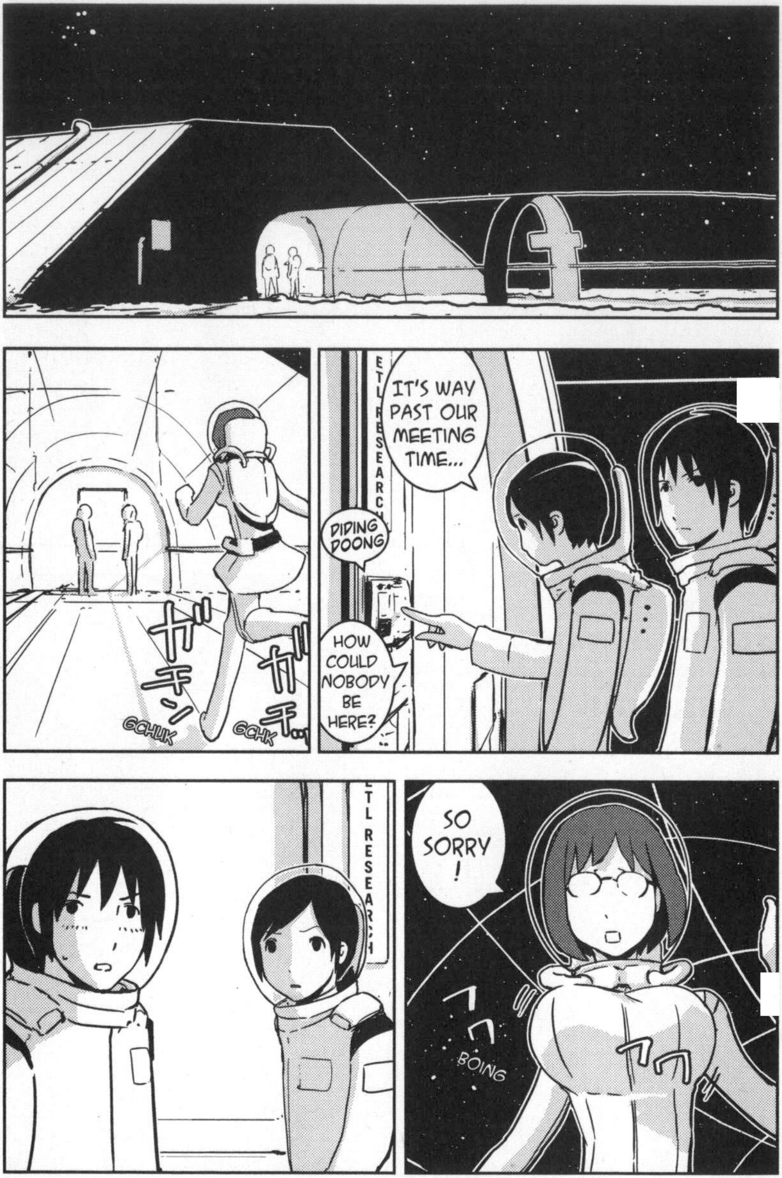
ETL RESEARCH
IT'S WAY PAST OUR MEETING TIME...
DIDING DOONG
HOW COULD NOBODY BE HERE?
GCHUK
GCHK
SO SORRY!
BOING
ETL RESEARCH

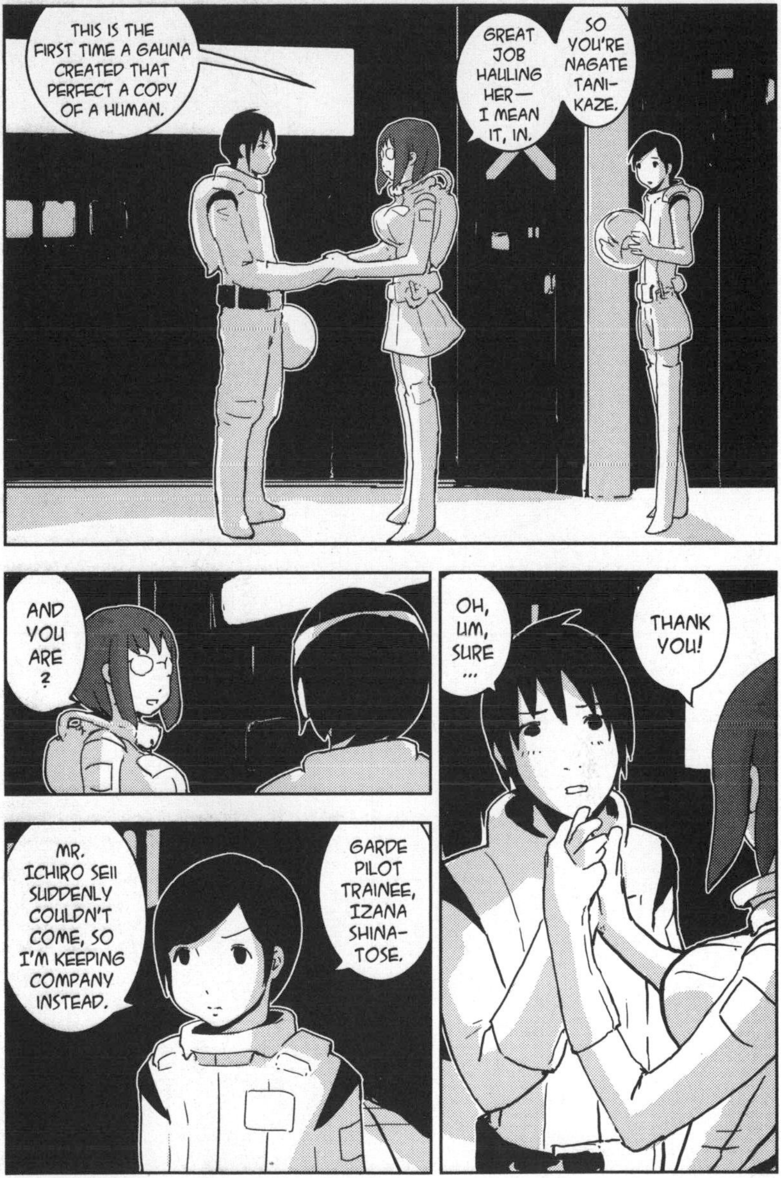
SO YOU'RE NAGATE TANI-KAZE.
GREAT JOB HAULING HER— I MEAN IT, IN.
THIS IS THE FIRST TIME A GAUNA CREATED THAT PERFECT A COPY OF A HUMAN.
THANK YOU!
OH, UM, SURE ...
AND YOU ARE?
GARDE PILOT TRAINEE, IZANA SHINA-TOSE.
MR. ICHIRO SEII SUDDENLY COULDN'T COME, SO I'M KEEPING COMPANY INSTEAD.

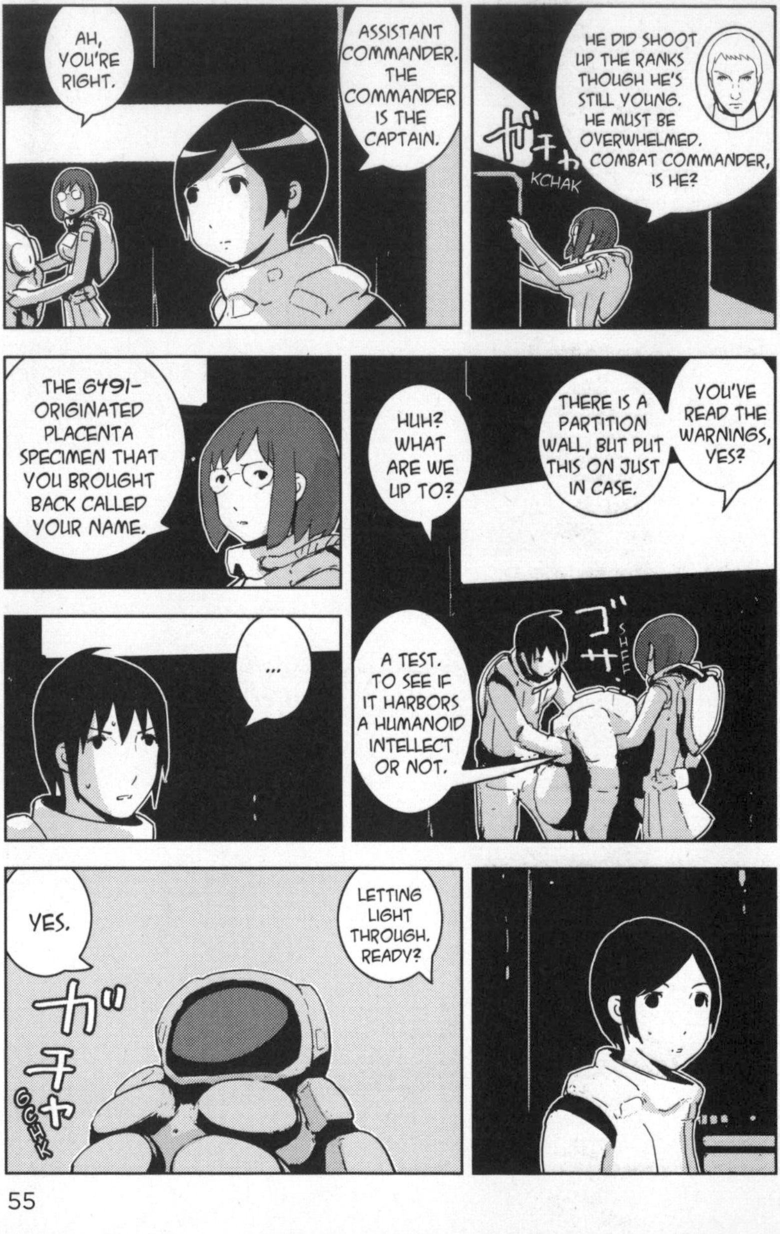
HE DID SHOOT UP THE RANKS THOUGH HE'S STILL YOUNG. HE MUST BE OVERWHELMED. COMBAT COMMANDER, IS HE?
ガチャ
KCHAK
ASSISTANT COMMANDER. THE COMMANDER IS THE CAPTAIN.
AH, YOU'RE RIGHT.
YOU'VE READ THE WARNINGS, YES?
THERE IS A PARTITION WALL, BUT PUT THIS ON JUST IN CASE.
HUH? WHAT ARE WE UP TO?
ゴサ
SHFF
A TEST. TO SEE IF IT HARBORS A HUMANOID INTELLECT OR NOT.
THE 6491-ORIGINATED PLACENTA SPECIMEN THAT YOU BROUGHT BACK CALLED YOUR NAME.
...
LETTING LIGHT THROUGH. READY?
YES.
ガチャ
GCHK

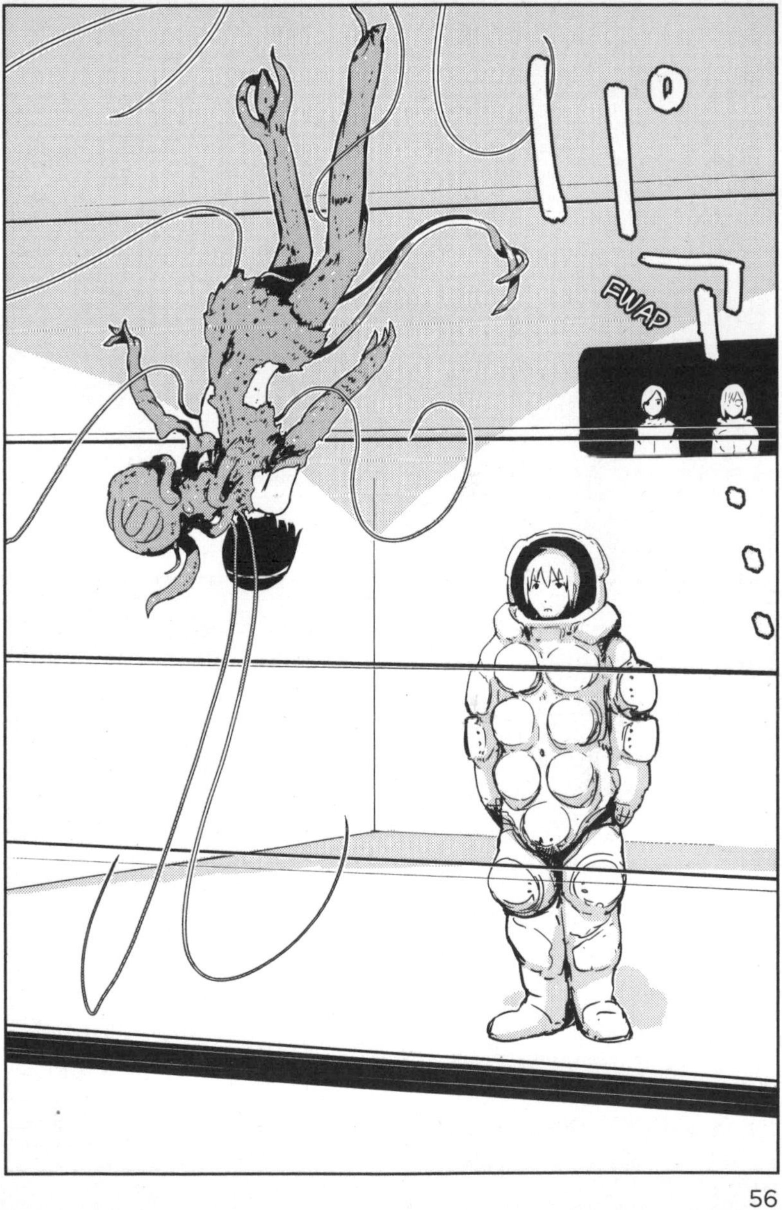
バ
FWAP

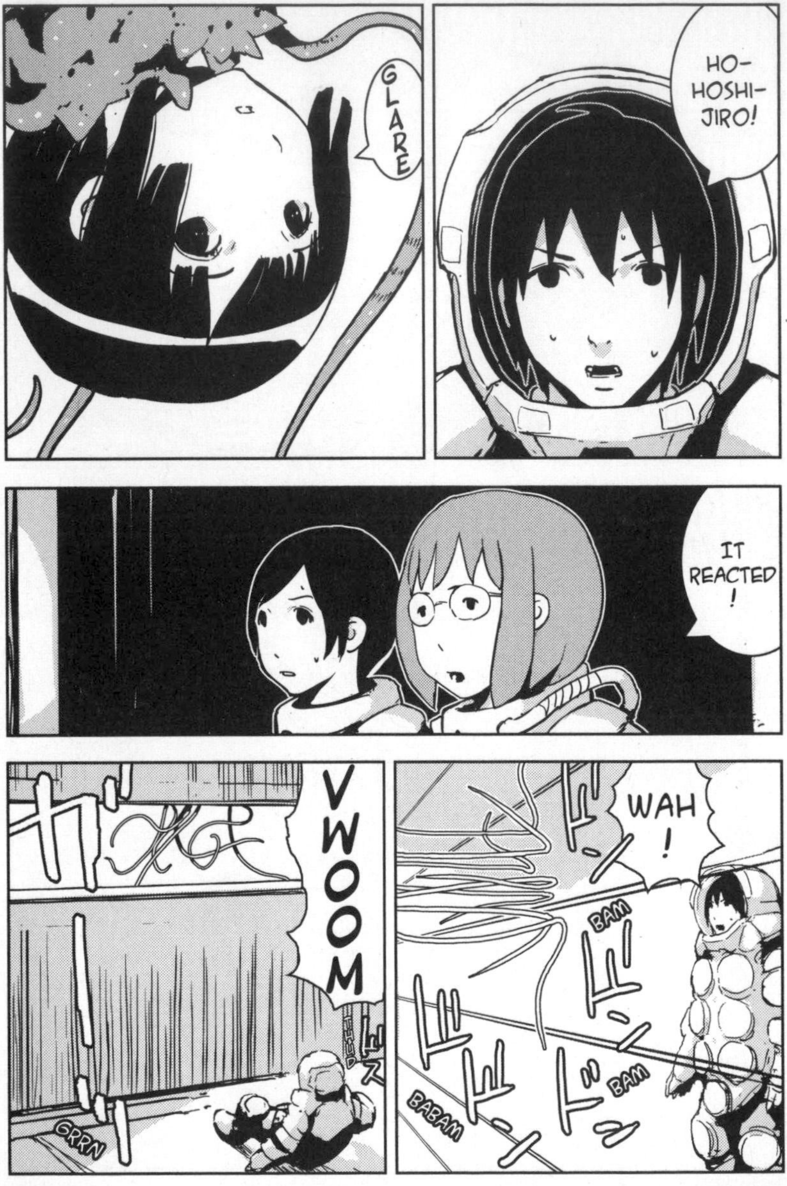
HO-HOSHI-JIRO!
GLARE
IT REACTED!
WAH!
BAM
BAM
BABAM
VWOOM
THUD
GRRN

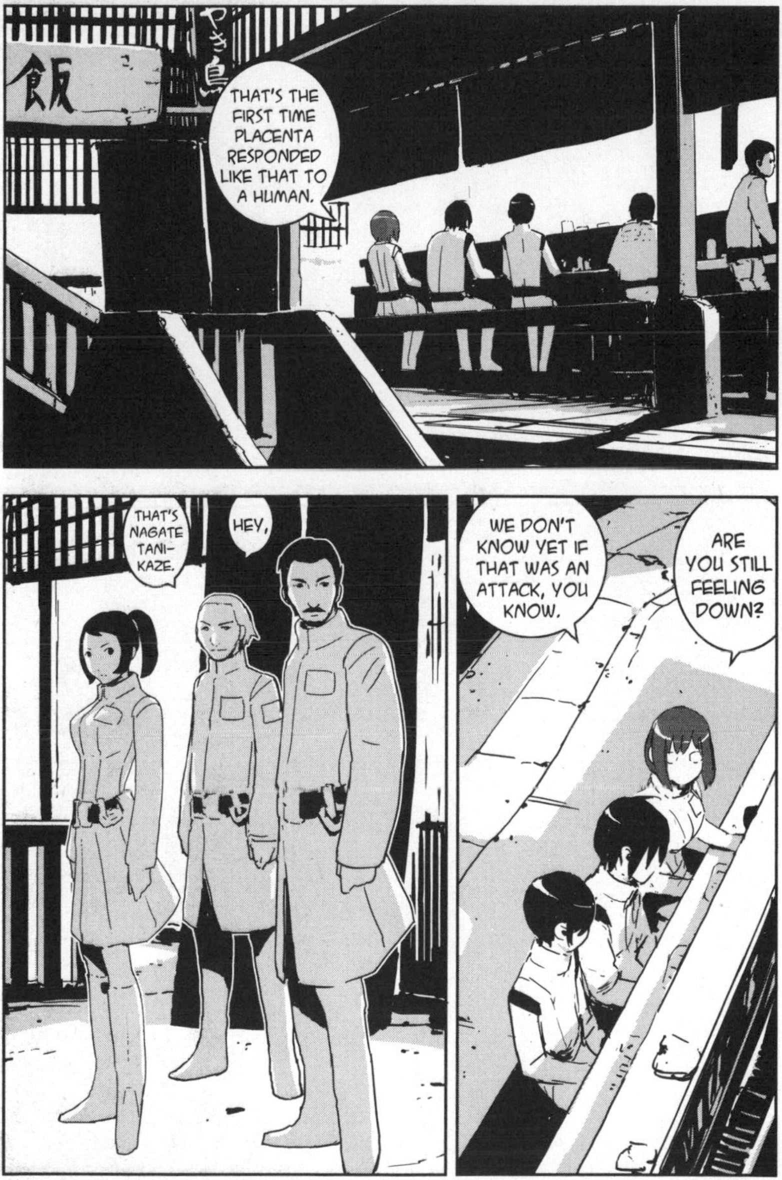
やき鳥
飯
THAT'S THE FIRST TIME PLACENTA RESPONDED LIKE THAT TO A HUMAN.
ARE YOU STILL FEELING DOWN?
WE DON'T KNOW YET IF THAT WAS AN ATTACK, YOU KNOW.
HEY,
THAT'S NAGATE TANI-KAZE.

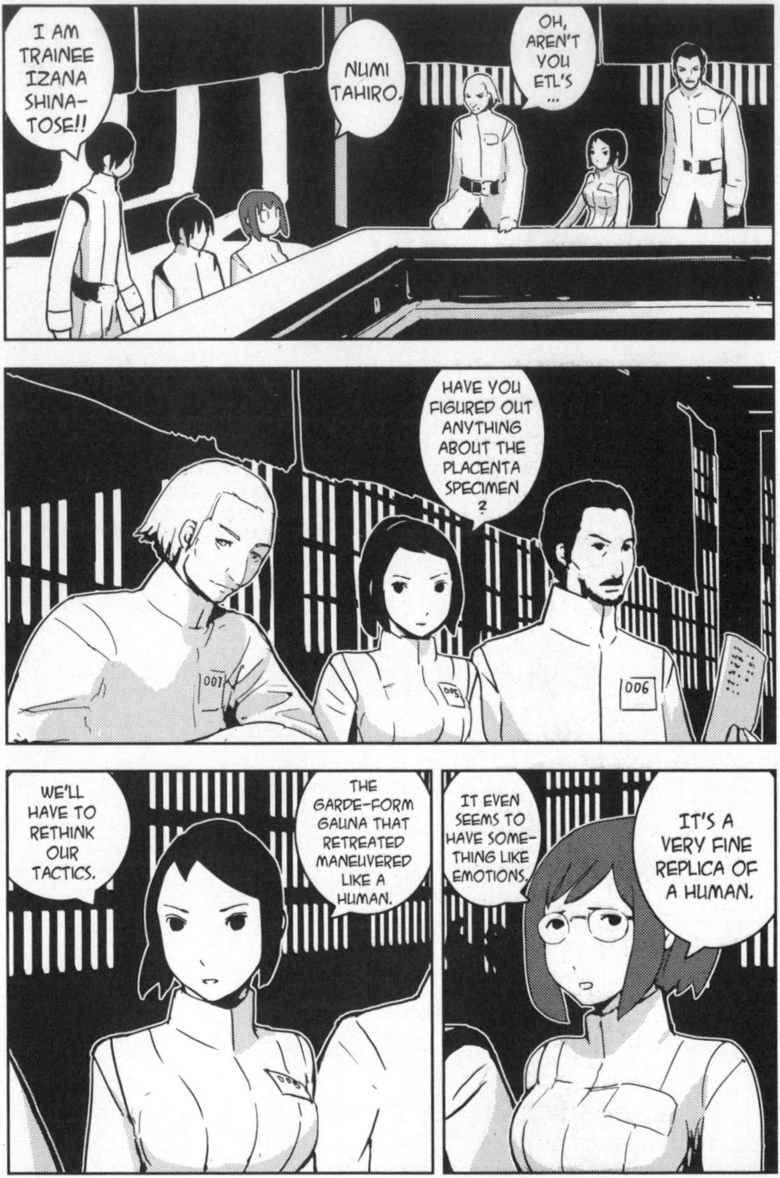
OH, AREN'T YOU ETL'S ...
NUMI TAHIRO.
I AM TRAINEE IZANA SHINA-TOSE!!
HAVE YOU FIGURED OUT ANYTHING ABOUT THE PLACENTA SPECIMEN ?
006
005
001
IT'S A VERY FINE REPLICA OF A HUMAN.
IT EVEN SEEMS TO HAVE SOME-THING LIKE EMOTIONS.
THE GARDE-FORM GAUNA THAT RETREATED MANEUVERED LIKE A HUMAN.
WE'LL HAVE TO RETHINK OUR TACTICS.

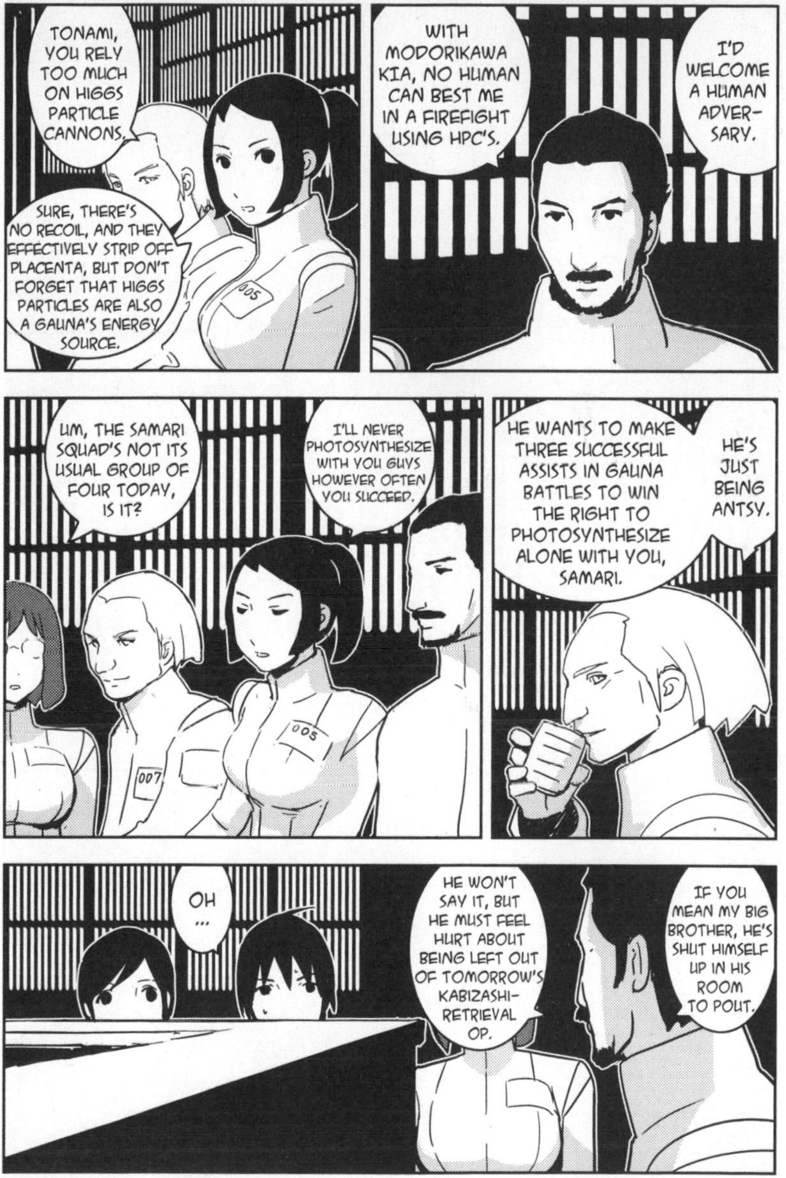
I'D WELCOME A HUMAN ADVER-SARY.
WITH MODORIKAWA KIA, NO HUMAN CAN BEST ME IN A FIREFIGHT USING HPC'S.
TONAMI, YOU RELY TOO MUCH ON HIGGS PARTICLE CANNONS.
SURE, THERE'S NO RECOIL, AND THEY EFFECTIVELY STRIP OFF PLACENTA, BUT DON'T FORGET THAT HIGGS PARTICLES ARE ALSO A GAUNA'S ENERGY SOURCE.
HE'S JUST BEING ANTSY.
HE WANTS TO MAKE THREE SUCCESSFUL ASSISTS IN GAUNA BATTLES TO WIN THE RIGHT TO PHOTOSYNTHESIZE ALONE WITH YOU, SAMARI.
I'LL NEVER PHOTOSYNTHESIZE WITH YOU GUYS HOWEVER OFTEN YOU SUCCEED.
UM, THE SAMARI SQUAD'S NOT ITS USUAL GROUP OF FOUR TODAY, IS IT?
IF YOU MEAN MY BIG BROTHER, HE'S SHUT HIMSELF UP IN HIS ROOM TO POUT.
HE WON'T SAY IT, BUT HE MUST FEEL HURT ABOUT BEING LEFT OUT OF TOMORROW'S KABIZASHI-RETRIEVAL OP.
OH ...

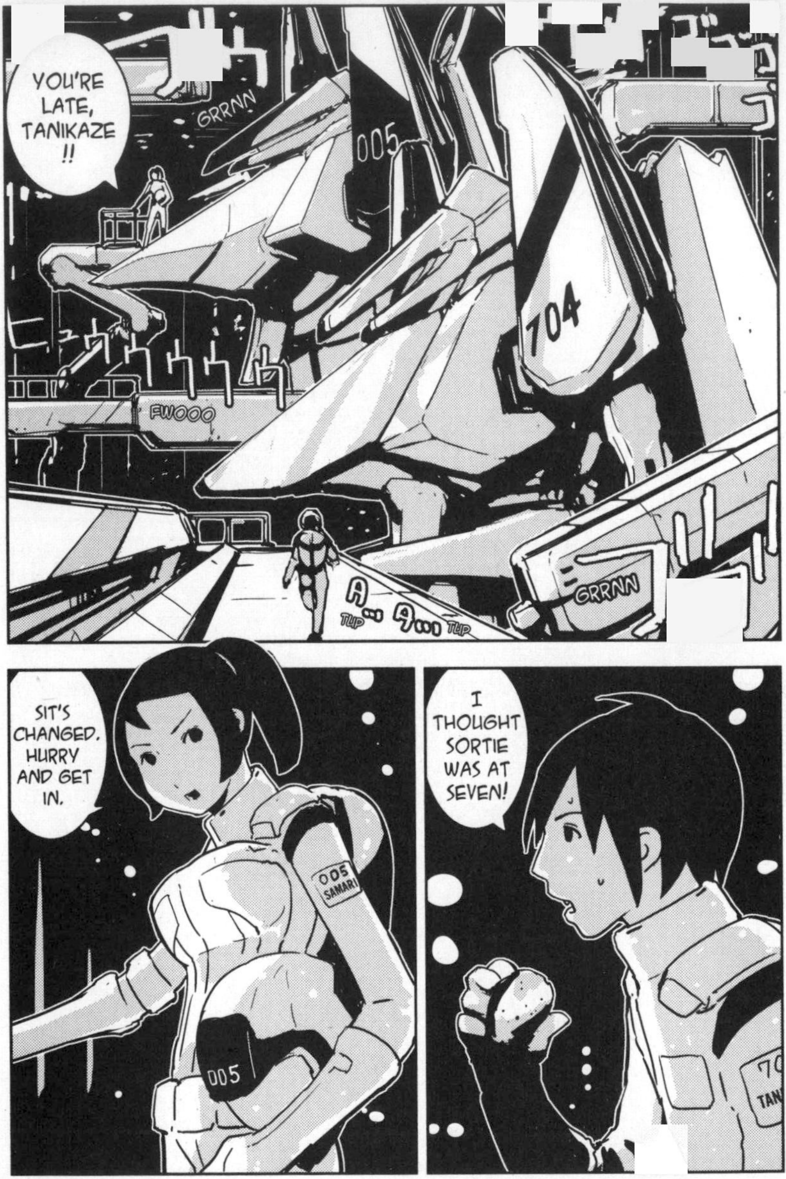
YOU'RE LATE, TANIKAZE !!
GRRNN
005
704
FWOOO
GRRNN
TUP TUP
I THOUGHT SORTIE WAS AT SEVEN!
SIT'S CHANGED. HURRY AND GET IN.
005 SAMARI
005

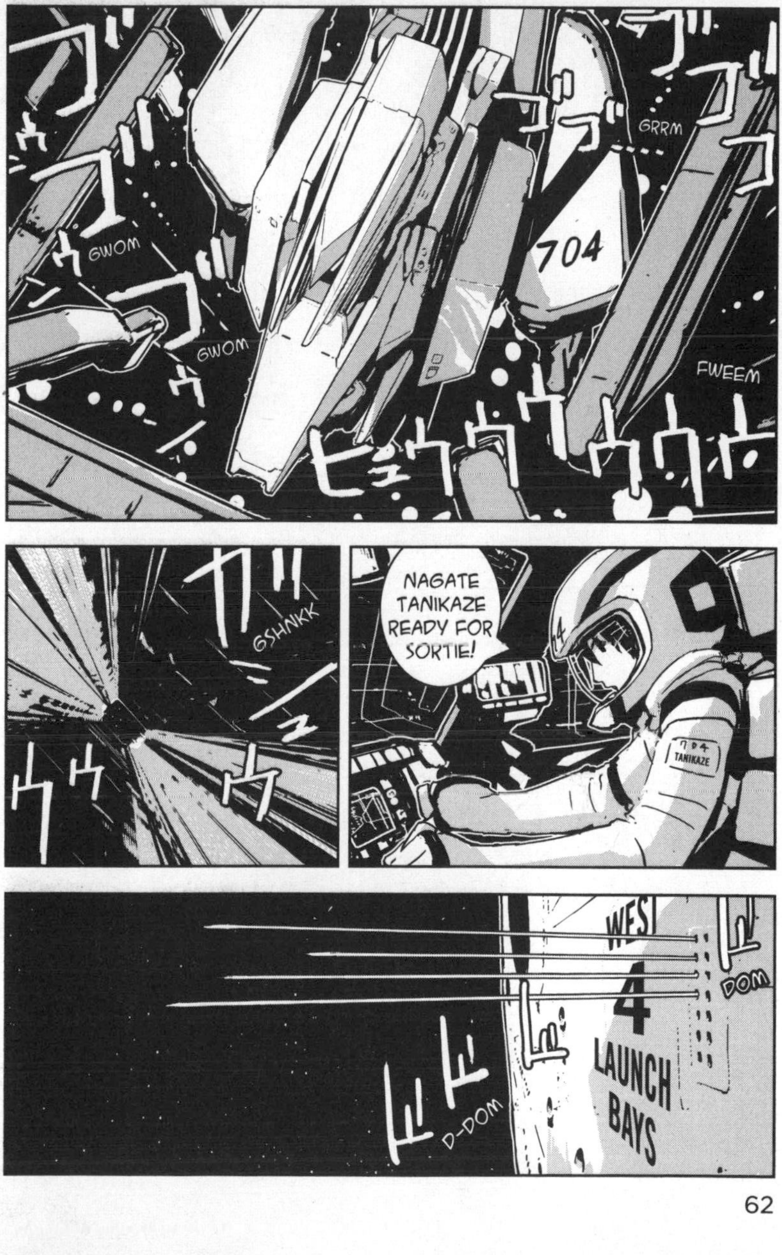
GRRM
704
FWEEM
GWOM
GWOM
NAGATE TANIKAZE READY FOR SORTIE!
704 TANIKAZE
GSHNKK
WEST 4 LAUNCH BAYS
DOM
D-DOM

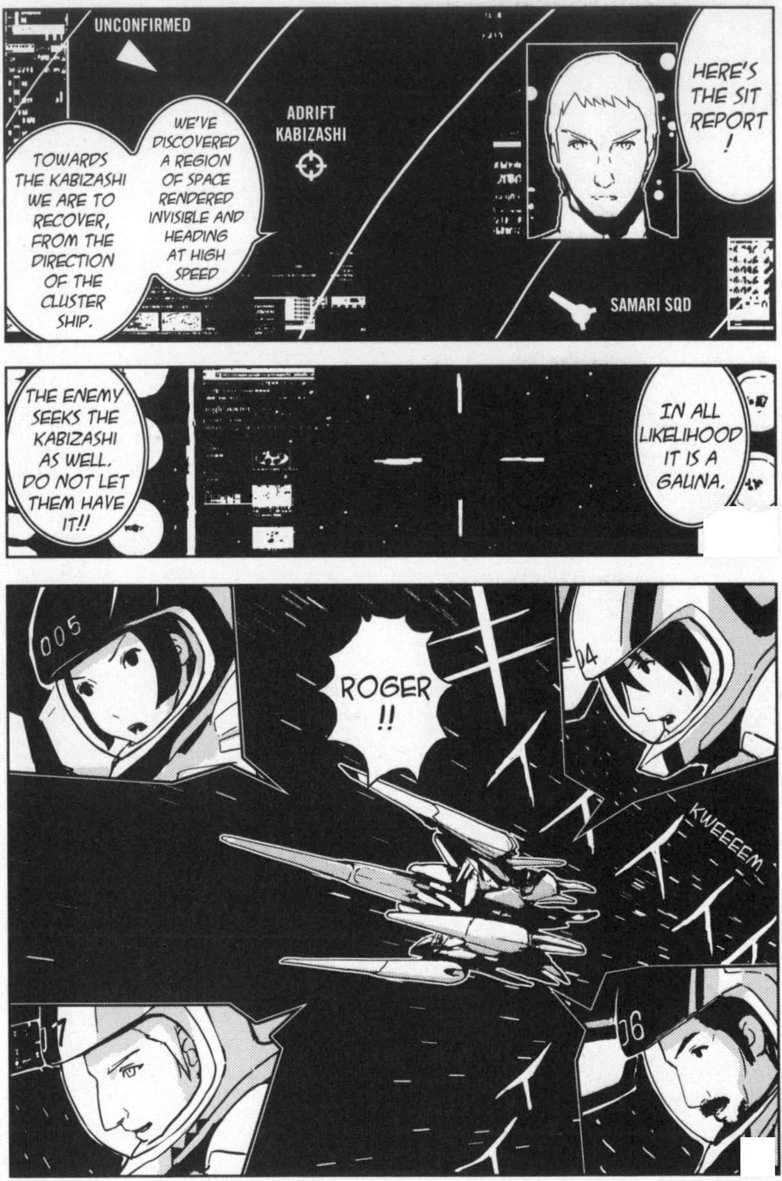
HERE'S THE SIT REPORT!
SAMARI SQD
ADRIFT KABIZASHI
UNCONFIRMED
WE'VE DISCOVERED A REGION OF SPACE RENDERED INVISIBLE AND HEADING AT HIGH SPEED
TOWARDS THE KABIZASHI WE ARE TO RECOVER, FROM THE DIRECTION OF THE CLUSTER SHIP.
IN ALL LIKELIHOOD IT IS A GAUNA.
THE ENEMY SEEKS THE KABIZASHI AS WELL. DO NOT LET THEM HAVE IT!!
04
ROGER!!
005
KWEEEEM
06
07

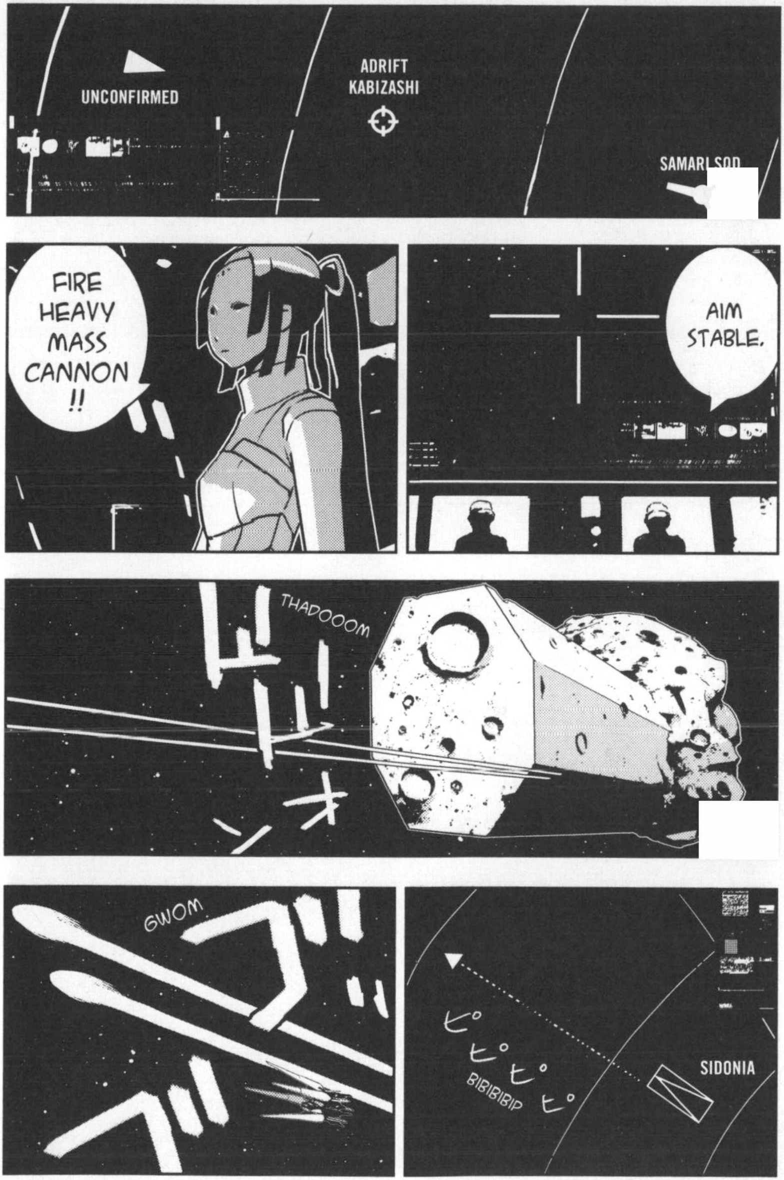

ADRIFT
KABIZASHI
UNCONFIRMED
SAMARI SOD
AIM STABLE.
FIRE HEAVY MASS CANNON !!
THADOOOM
SIDONIA
BIBIBIBIP
GWOM

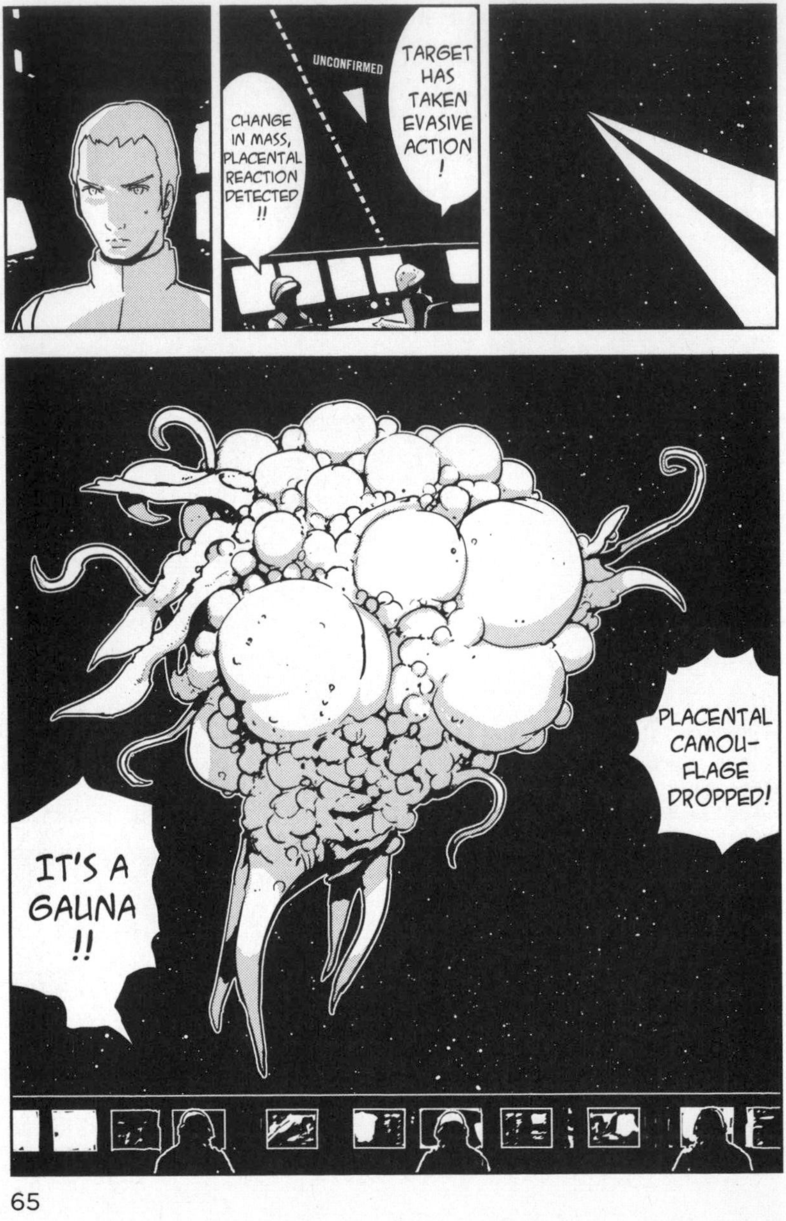
TARGET HAS TAKEN EVASIVE ACTION!
UNCONFIRMED
CHANGE IN MASS, PLACENTAL REACTION DETECTED!!
PLACENTAL CAMOU-FLAGE DROPPED!
IT'S A GAUNA!!

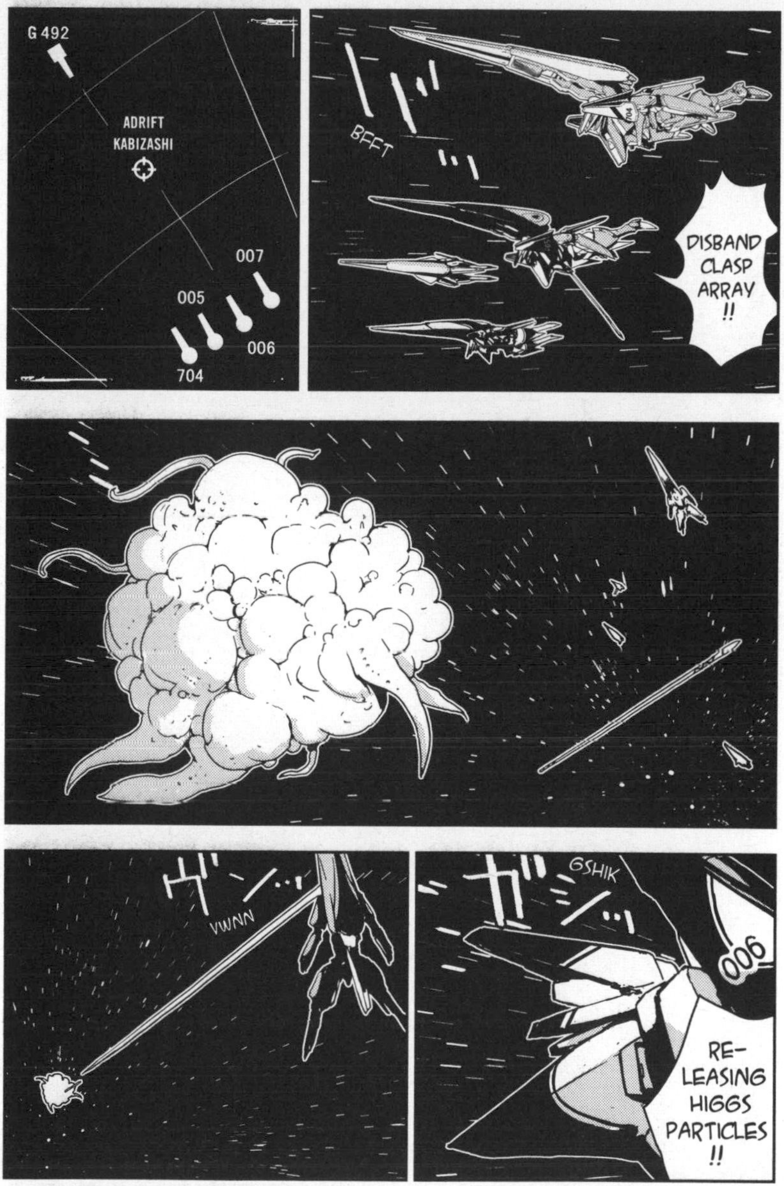
BFFT
DISBAND CLASP ARRAY !!
G 492
ADRIFT KABIZASHI
007
005
006
704
GSHIK
006
RE-LEASING HIGGS PARTICLES !!
VWNN

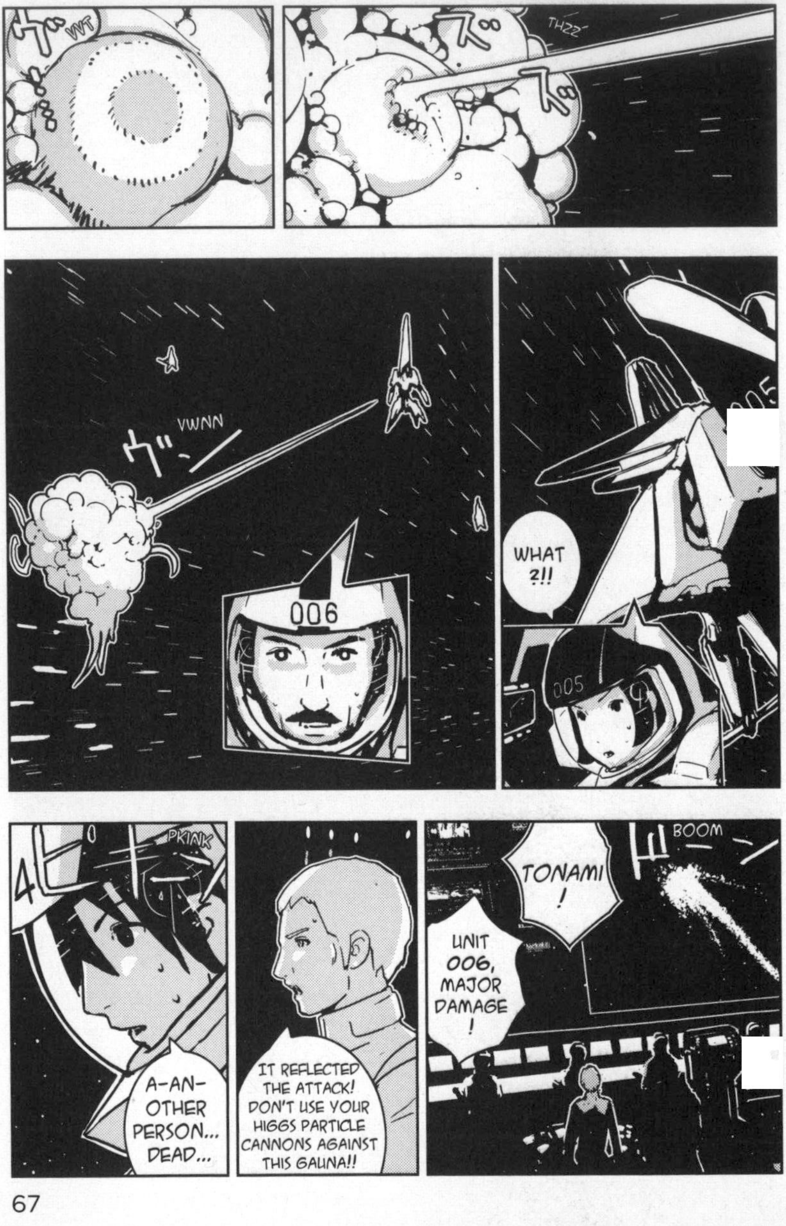
THZZ
VVT
WHAT ?!!
VWNN
BOOM
TONAMI !
UNIT 006, MAJOR DAMAGE !
IT REFLECTED THE ATTACK! DON'T USE YOUR HIGGS PARTICLE CANNONS AGAINST THIS GAUNA!!
PKINK
A-AN-OTHER PERSON... DEAD...

005
G 492
KABIZASHI
007
704
ZLRRT
GWOP
THOOM
UH-UH !!
007
007
KANK
CHOMP
FWING
FWING
CATCH
KABIZASHI RETRIEVED !!

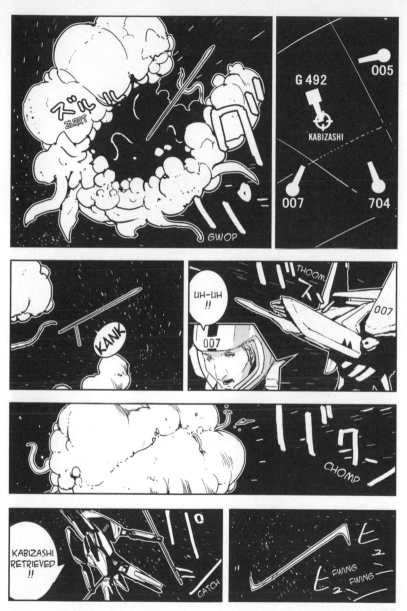

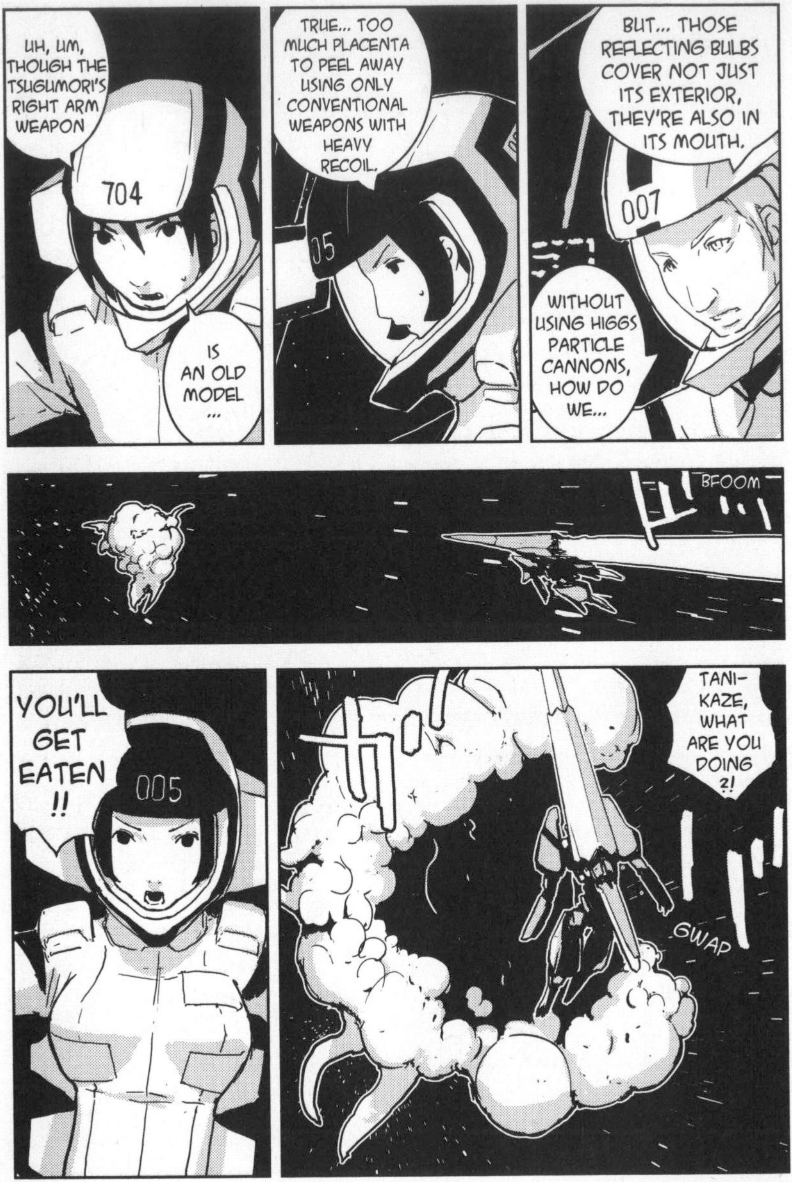
BUT... THOSE REFLECTING BULBS COVER NOT JUST ITS EXTERIOR, THEY'RE ALSO IN ITS MOUTH.
007
WITHOUT USING HIGGS PARTICLE CANNONS, HOW DO WE...
TRUE... TOO MUCH PLACENTA TO PEEL AWAY USING ONLY CONVENTIONAL WEAPONS WITH HEAVY RECOIL.
05
UH, UM, THOUGH THE TSUGUMORI'S RIGHT ARM WEAPON
704
IS AN OLD MODEL ...
BFOOM
TANI-KAZE, WHAT ARE YOU DOING ?!
GWAP
YOU'LL GET EATEN !!
005

GLOMP
CREAK
GGT
GRAB
CREAK
704
KCHK

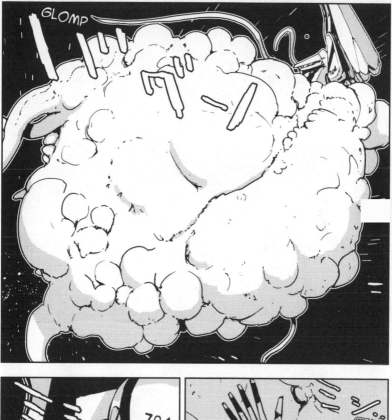

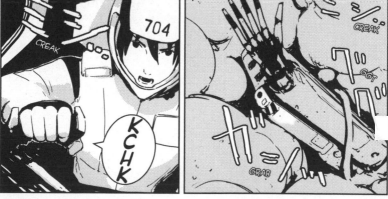

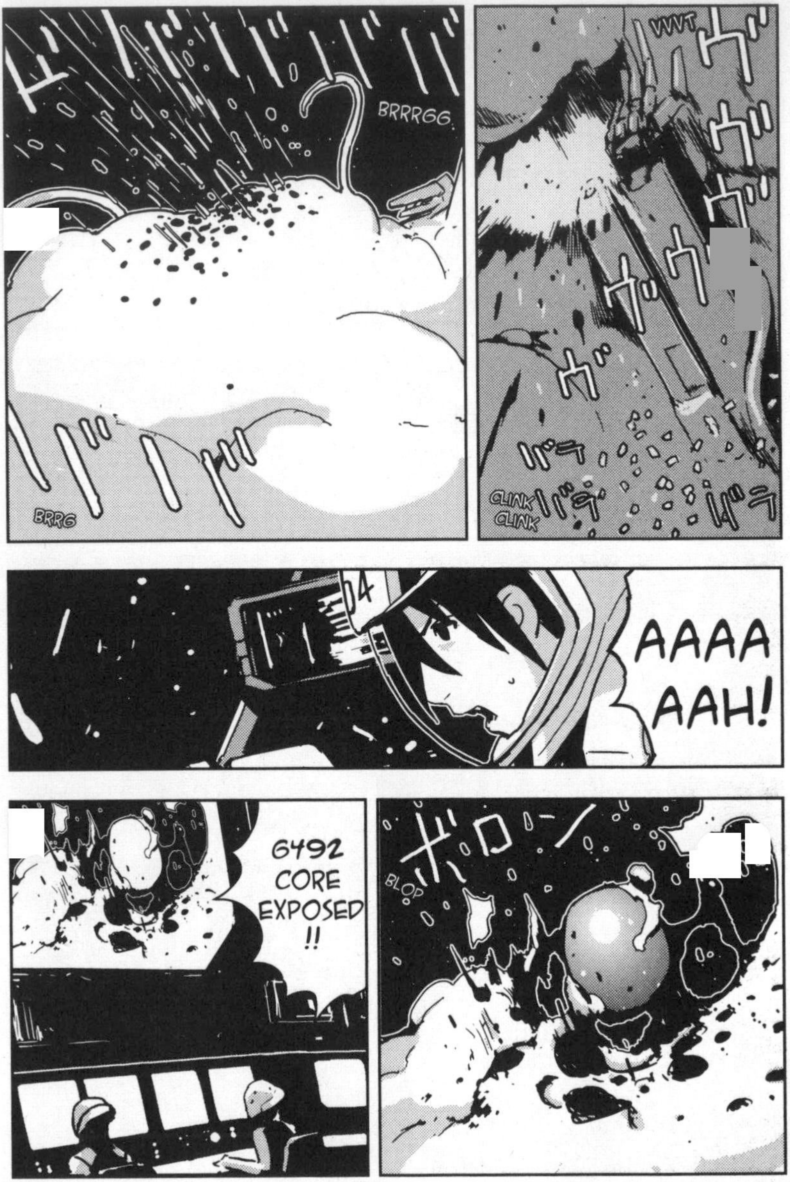
VVVT
CLINK CLINK
BRRRGG
BRRG
AAAA AAH!
BLOP
6492 CORE EXPOSED !!

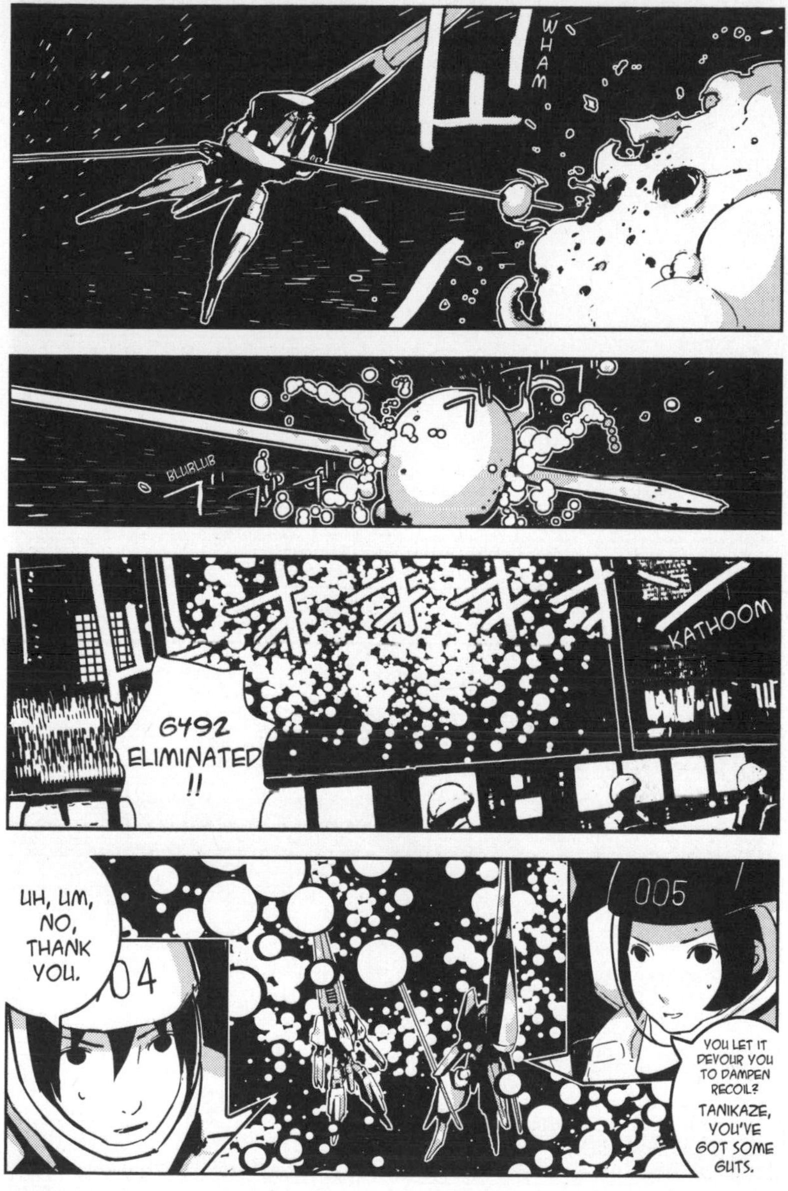
WHAM
BLUBLUB
KATHOOM
6492 ELIMINATED !!
005
YOU LET IT DEVOUR YOU TO DAMPEN RECOIL? TANIKAZE, YOU'VE GOT SOME GUTS.
UH, UM, NO, THANK YOU.

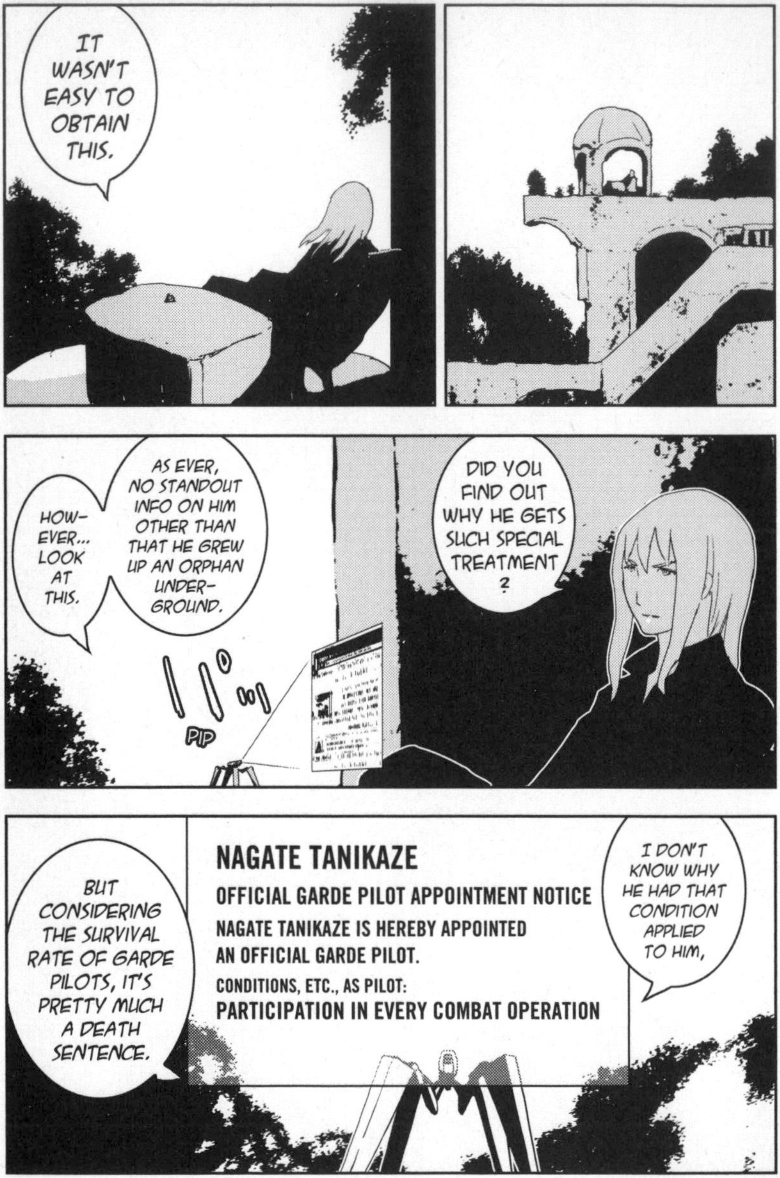
IT WASN'T EASY TO OBTAIN THIS.
DID YOU FIND OUT WHY HE GETS SUCH SPECIAL TREATMENT?
AS EVER, NO STANDOUT INFO ON HIM OTHER THAN THAT HE GREW UP AN ORPHAN UNDER-GROUND.
HOW-EVER... LOOK AT THIS.
PIP
I DON'T KNOW WHY HE HAD THAT CONDITION APPLIED TO HIM,
NAGATE TANIKAZE
OFFICIAL GARDE PILOT APPOINTMENT NOTICE
NAGATE TANIKAZE IS HEREBY APPOINTED AN OFFICIAL GARDE PILOT.
CONDITIONS, ETC., AS PILOT:
PARTICIPATION IN EVERY COMBAT OPERATION
BUT CONSIDERING THE SURVIVAL RATE OF GARDE PILOTS, IT'S PRETTY MUCH A DEATH SENTENCE.

Chapter 12: END

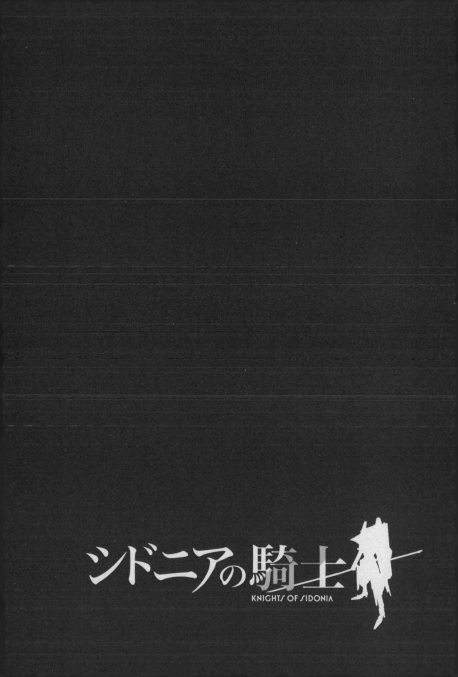

シドニアの騎士
KNIGHTS OF SIDONIA

Chapter 13: Izana Shinatose's Sigh

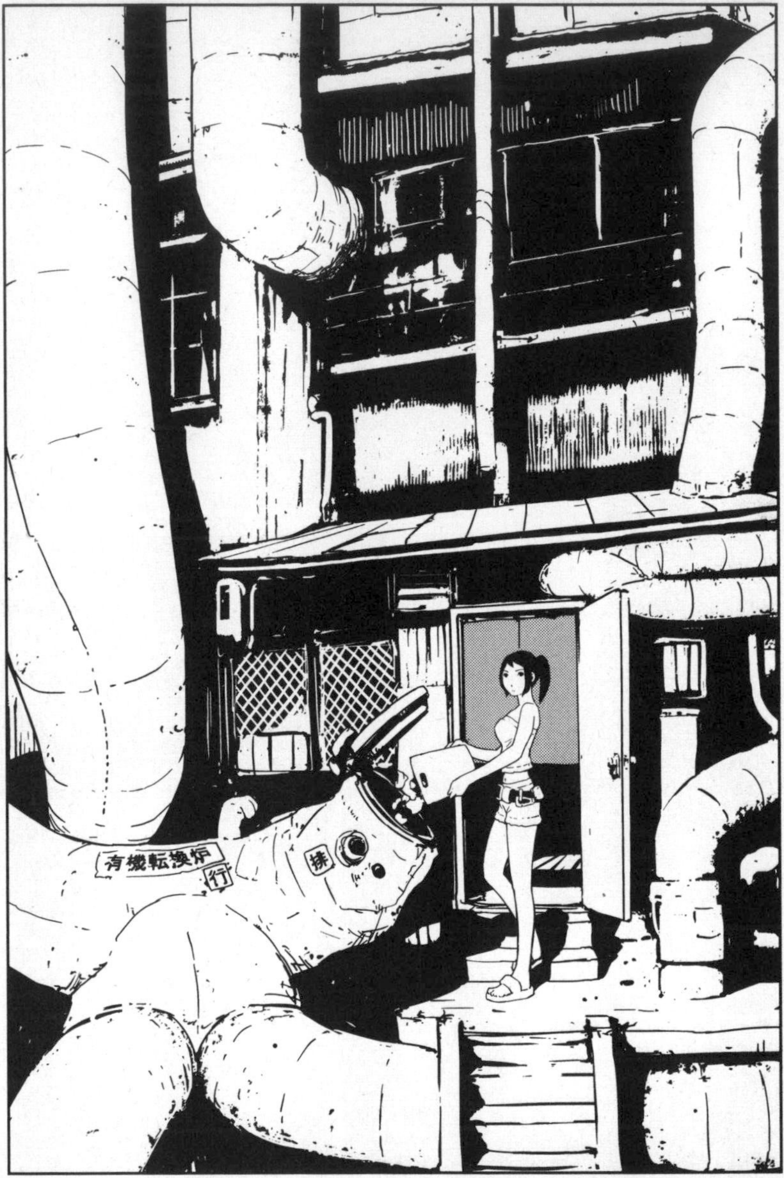
有機転換炉
行
排

One Hundred Sights of Sidonia Part Ten:
Organic-Conversion Reactor Residential Use Hatch

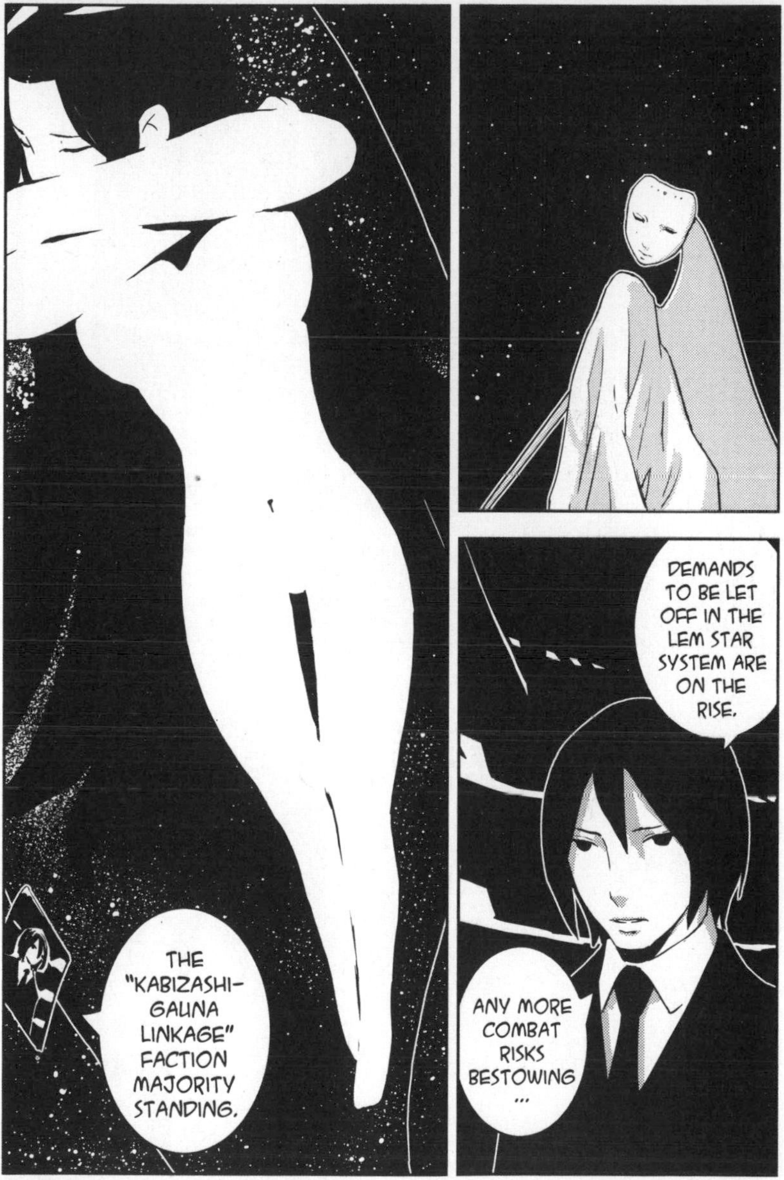
DEMANDS TO BE LET OFF IN THE LEM STAR SYSTEM ARE ON THE RISE.
ANY MORE COMBAT RISKS BESTOWING ...
THE "KABIZASHI-GAUNA LINKAGE" FACTION MAJORITY STANDING.

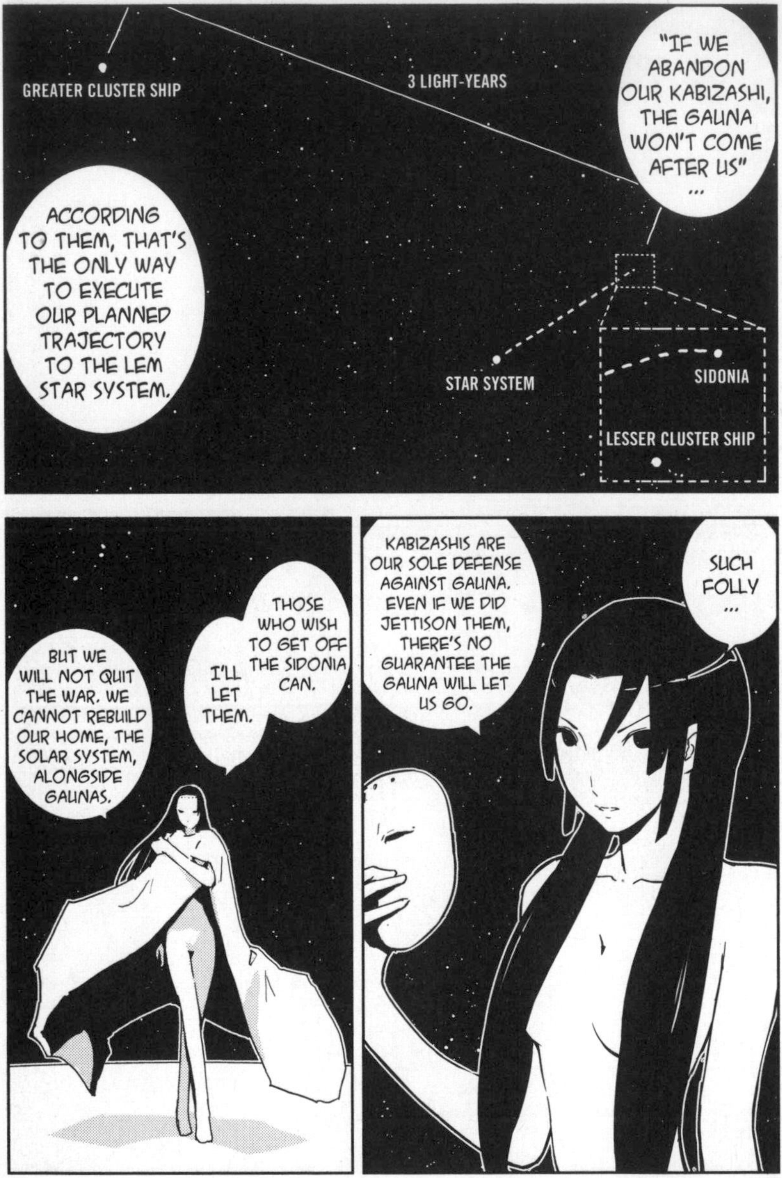
"IF WE ABANDON OUR KABIZASHI, THE GAUNA WON'T COME AFTER US" ...
GREATER CLUSTER SHIP
3 LIGHT-YEARS
ACCORDING TO THEM, THAT'S THE ONLY WAY TO EXECUTE OUR PLANNED TRAJECTORY TO THE LEM STAR SYSTEM.
SIDONIA
STAR SYSTEM
LESSER CLUSTER SHIP
SUCH FOLLY ...
KABIZASHIS ARE OUR SOLE DEFENSE AGAINST GAUNA. EVEN IF WE DID JETTISON THEM, THERE'S NO GUARANTEE THE GAUNA WILL LET US GO.
THOSE WHO WISH TO GET OFF THE SIDONIA CAN.
I'LL LET THEM.
BUT WE WILL NOT QUIT THE WAR. WE CANNOT REBUILD OUR HOME, THE SOLAR SYSTEM, ALONGSIDE GAUNAS.

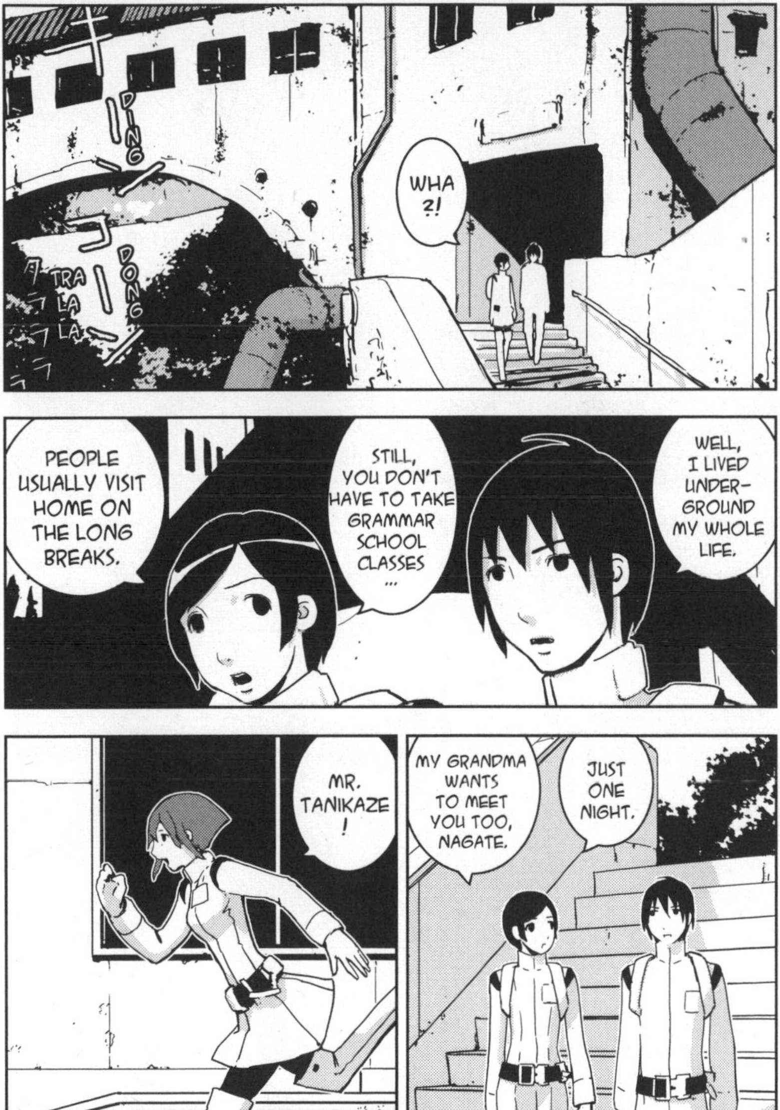
WHA ?!
DING DONG
TRA LA LA
WELL, I LIVED UNDER-GROUND MY WHOLE LIFE.
STILL, YOU DON'T HAVE TO TAKE GRAMMAR SCHOOL CLASSES ...
PEOPLE USUALLY VISIT HOME ON THE LONG BREAKS.
JUST ONE NIGHT.
MY GRANDMA WANTS TO MEET YOU TOO, NAGATE.
MR. TANIKAZE!

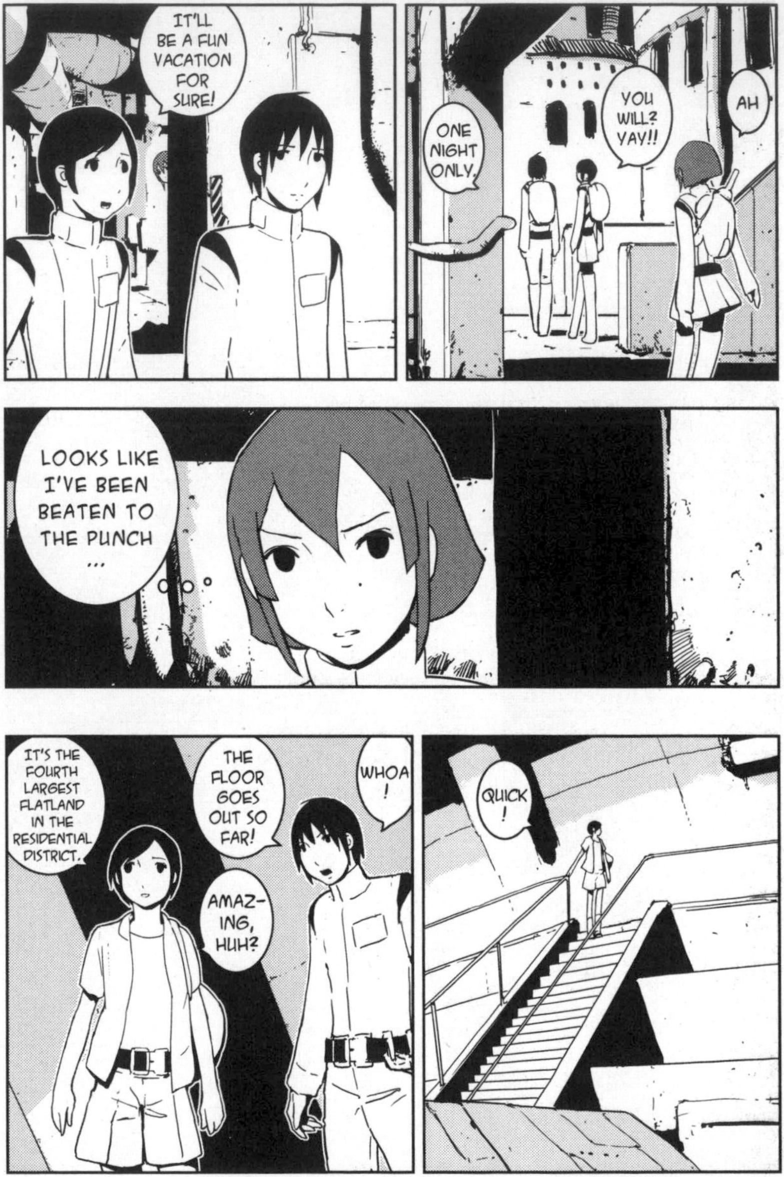
AH
YOU WILL? YAY!!
ONE NIGHT ONLY.
IT'LL BE A FUN VACATION FOR SURE!
LOOKS LIKE I'VE BEEN BEATEN TO THE PUNCH ...
QUICK!
WHOA!
THE FLOOR GOES OUT SO FAR!
AMAZ-ING, HUH?
IT'S THE FOURTH LARGEST FLATLAND IN THE RESIDENTIAL DISTRICT.

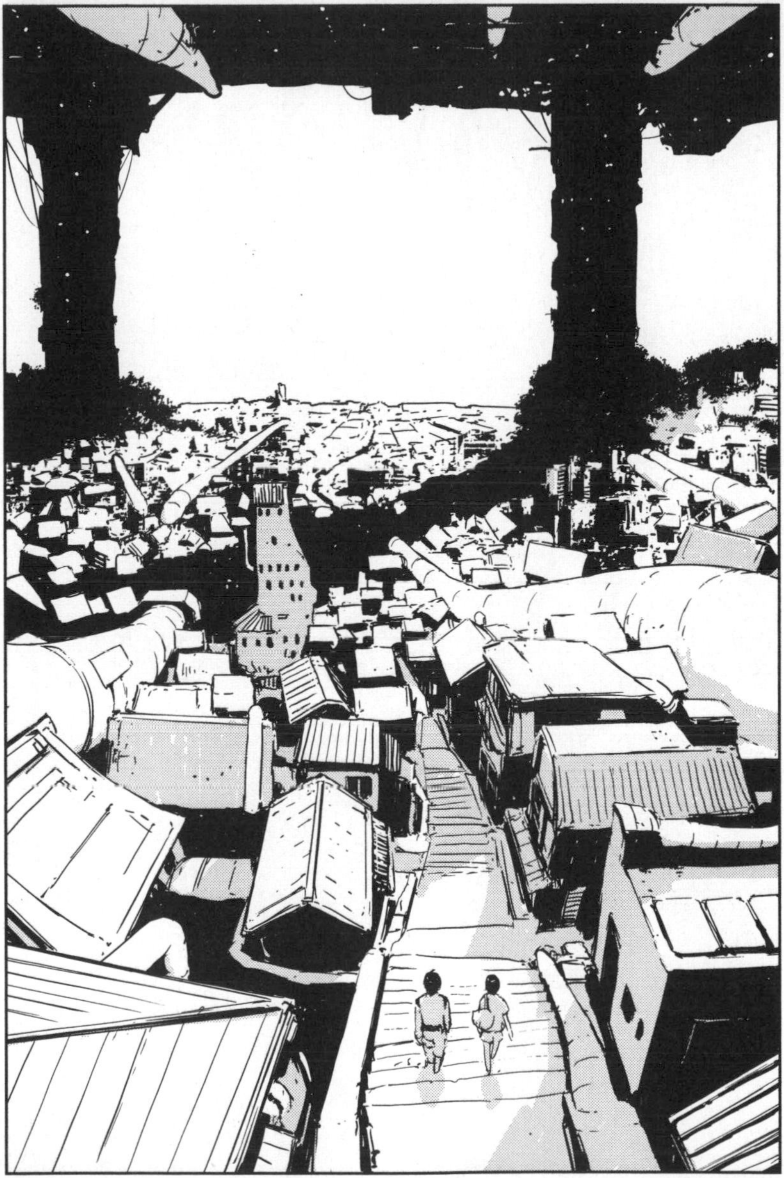

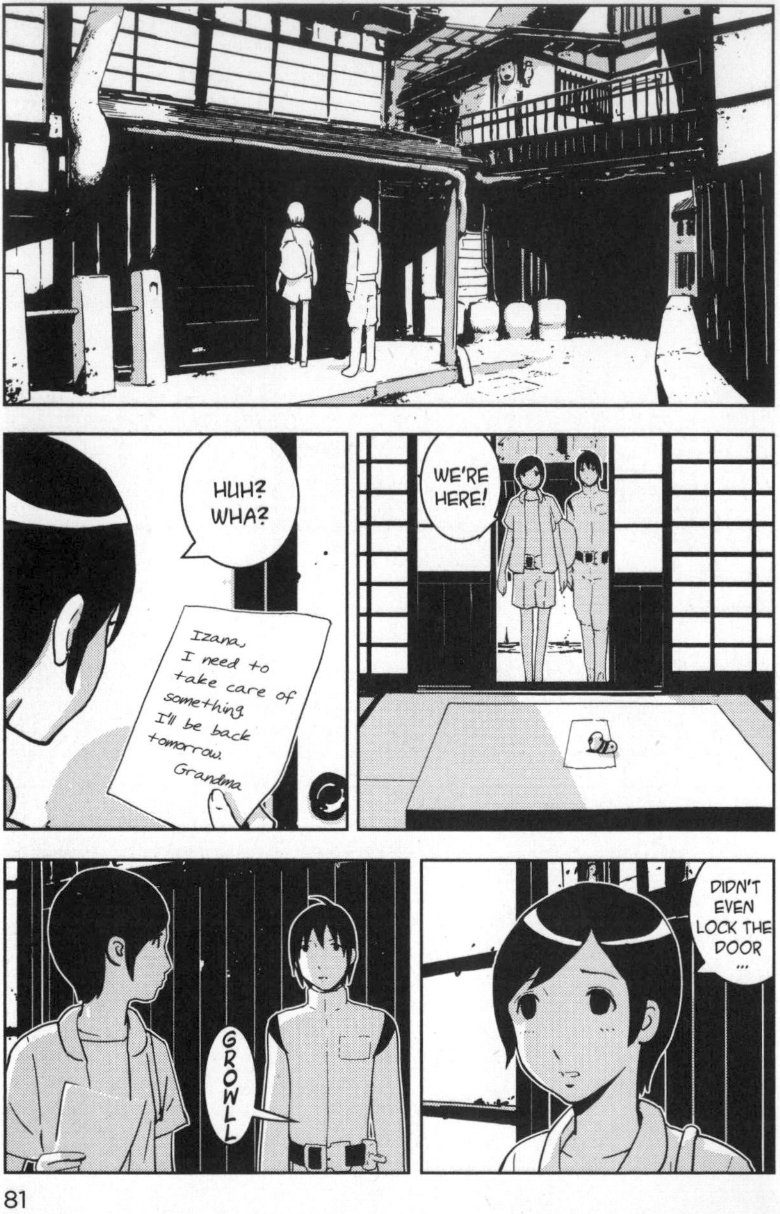
WE'RE HERE!
HUH? WHA?
Izana,
I need to take care of something.
I'll be back tomorrow.
Grandma
DIDN'T EVEN LOCK THE DOOR ...
GROWLL

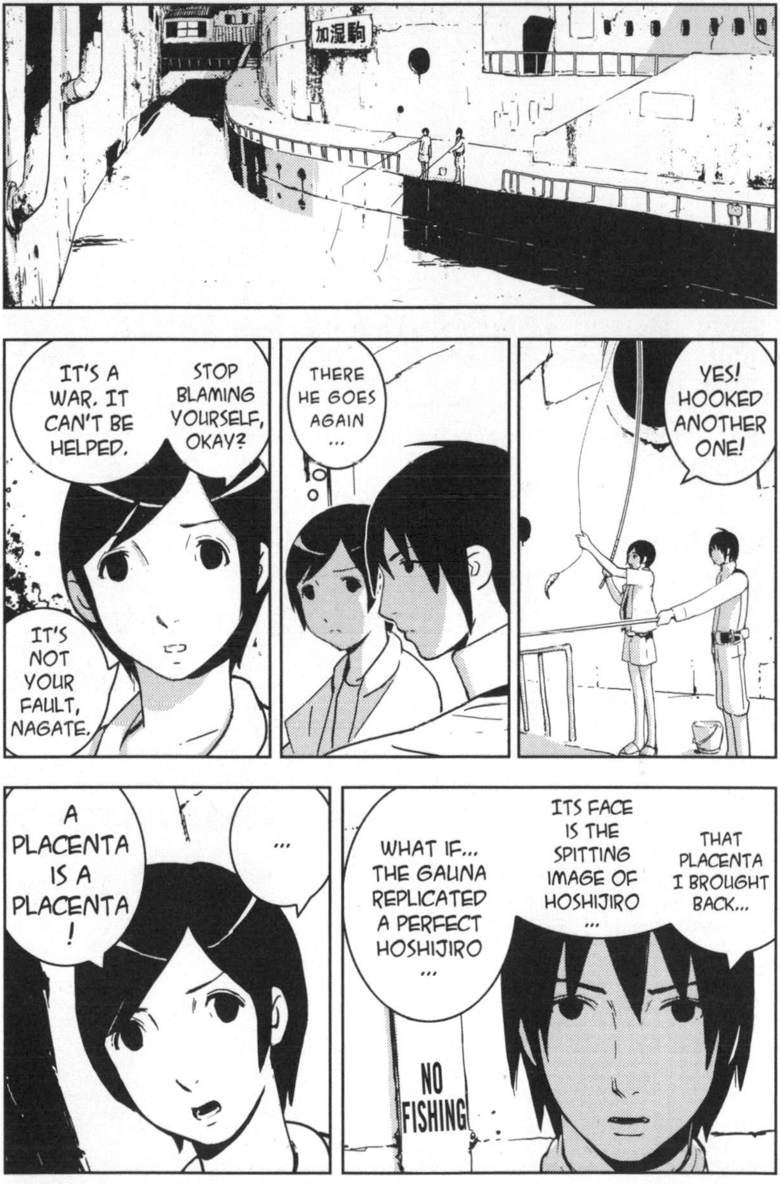
加湿駒
YES! HOOKED ANOTHER ONE!
THERE HE GOES AGAIN ...
STOP BLAMING YOURSELF, OKAY?
IT'S A WAR. IT CAN'T BE HELPED.
IT'S NOT YOUR FAULT, NAGATE.
THAT PLACENTA I BROUGHT BACK...
ITS FACE IS THE SPITTING IMAGE OF HOSHIJIRO ...
WHAT IF... THE GAUNA REPLICATED A PERFECT HOSHIJIRO ...
NO FISHING
...
A PLACENTA IS A PLACENTA !

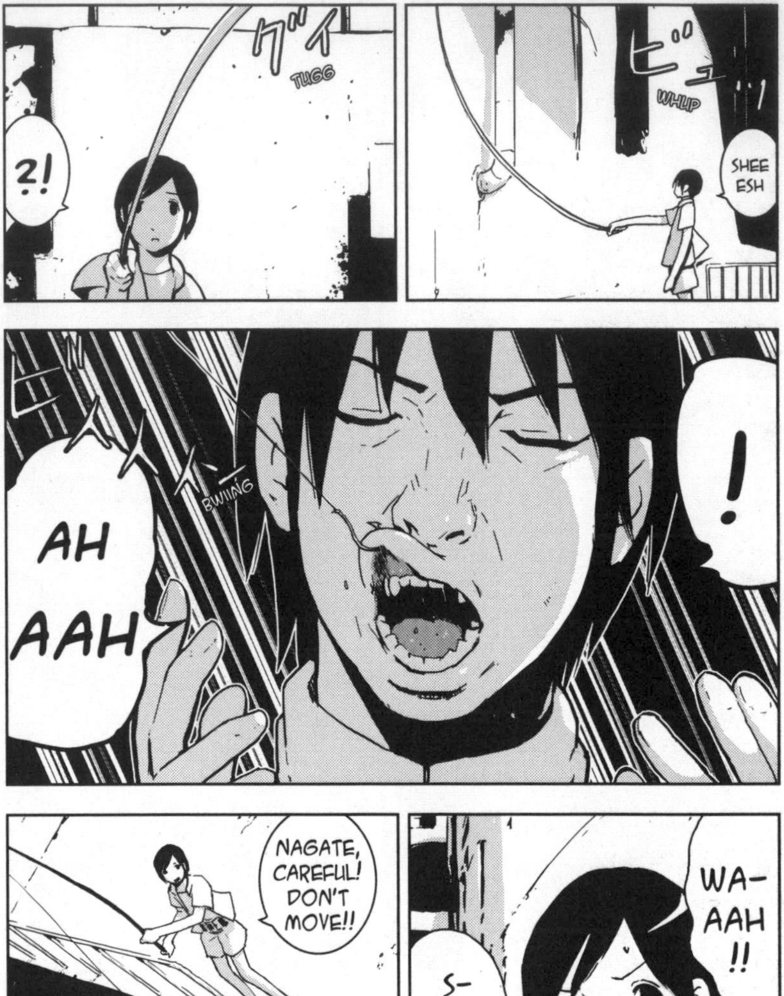
WHUP
SHEE ESH
TUGG
?!
!
BWIING
AH AAH
WA-AAH!!
S-SORRY! SORRY!!
NAGATE, CAREFUL! DON'T MOVE!!
YOW OWOW

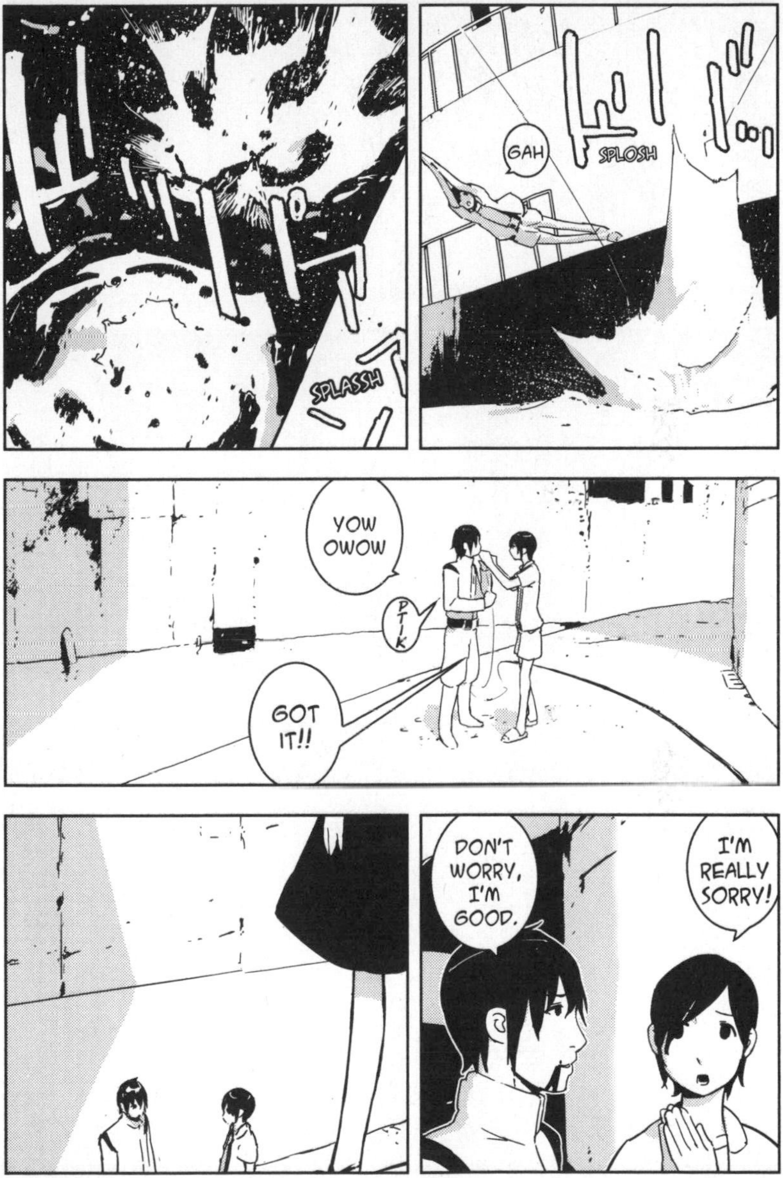
GAH
SPLOSH
SPLASSH
YOW OWOW
PTIK
GOT IT!!
I'M REALLY SORRY!
DON'T WORRY, I'M GOOD.

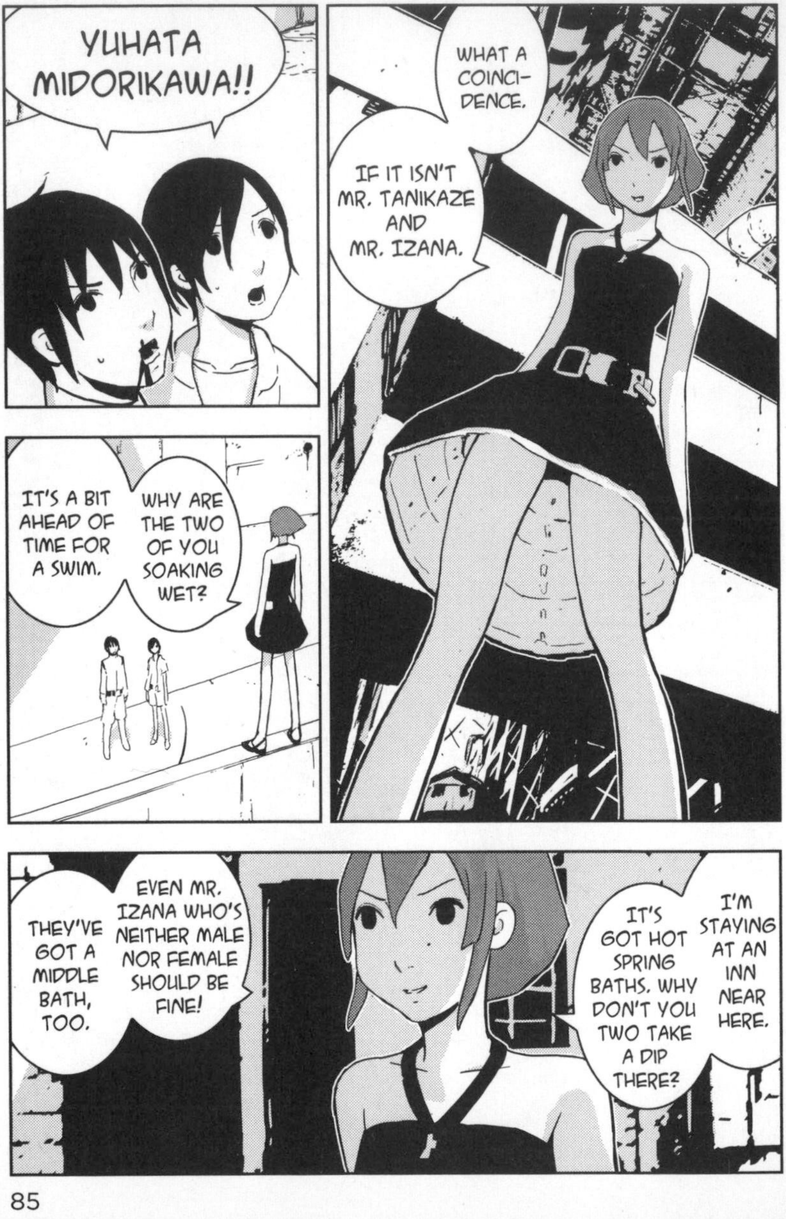
WHAT A COINCI-DENCE.
IF IT ISN'T MR. TANIKAZE AND MR. IZANA.
YUHATA MIDORIKAWA!!
WHY ARE THE TWO OF YOU SOAKING WET?
IT'S A BIT AHEAD OF TIME FOR A SWIM.
I'M STAYING AT AN INN NEAR HERE.
IT'S GOT HOT SPRING BATHS. WHY DON'T YOU TWO TAKE A DIP THERE?
EVEN MR. IZANA WHO'S NEITHER MALE NOR FEMALE SHOULD BE FINE!
THEY'VE GOT A MIDDLE BATH, TOO.

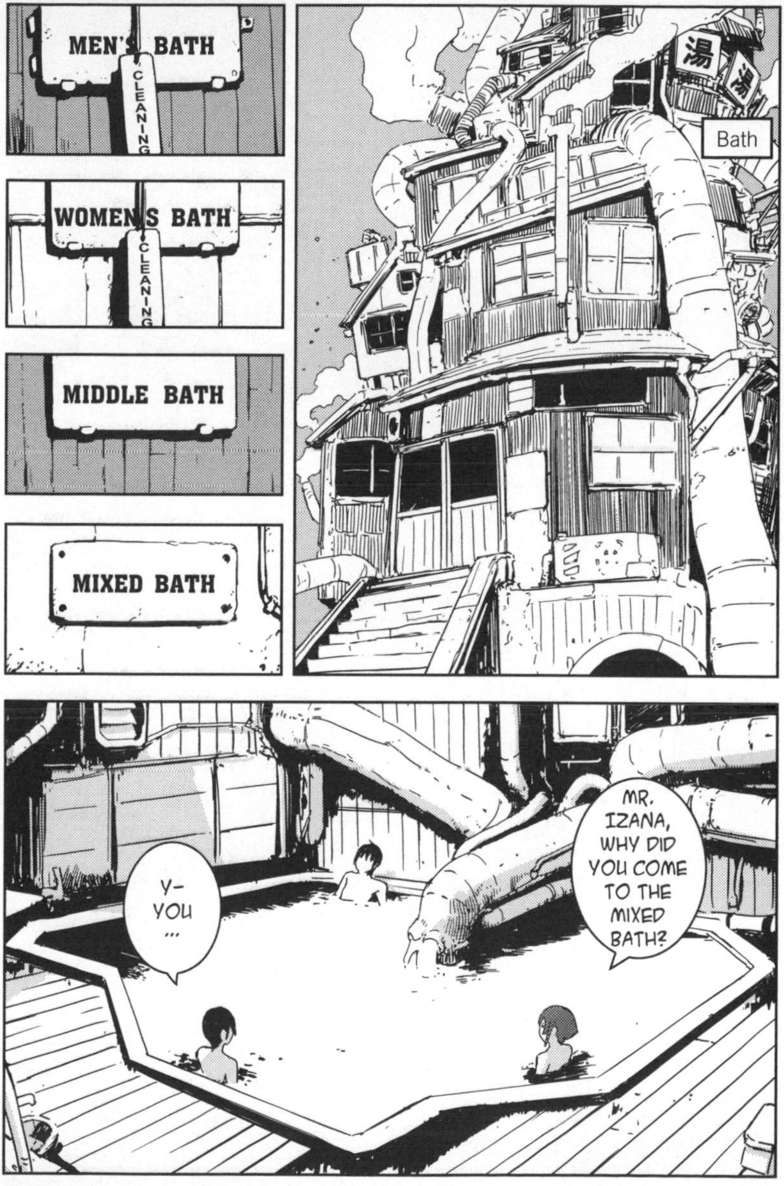
湯
湯
Bath
MEN'S BATH
CLEANING
WOMEN'S BATH
CLEANING
MIDDLE BATH
MIXED BATH
MR. IZANA, WHY DID YOU COME TO THE MIXED BATH?
Y-YOU ...

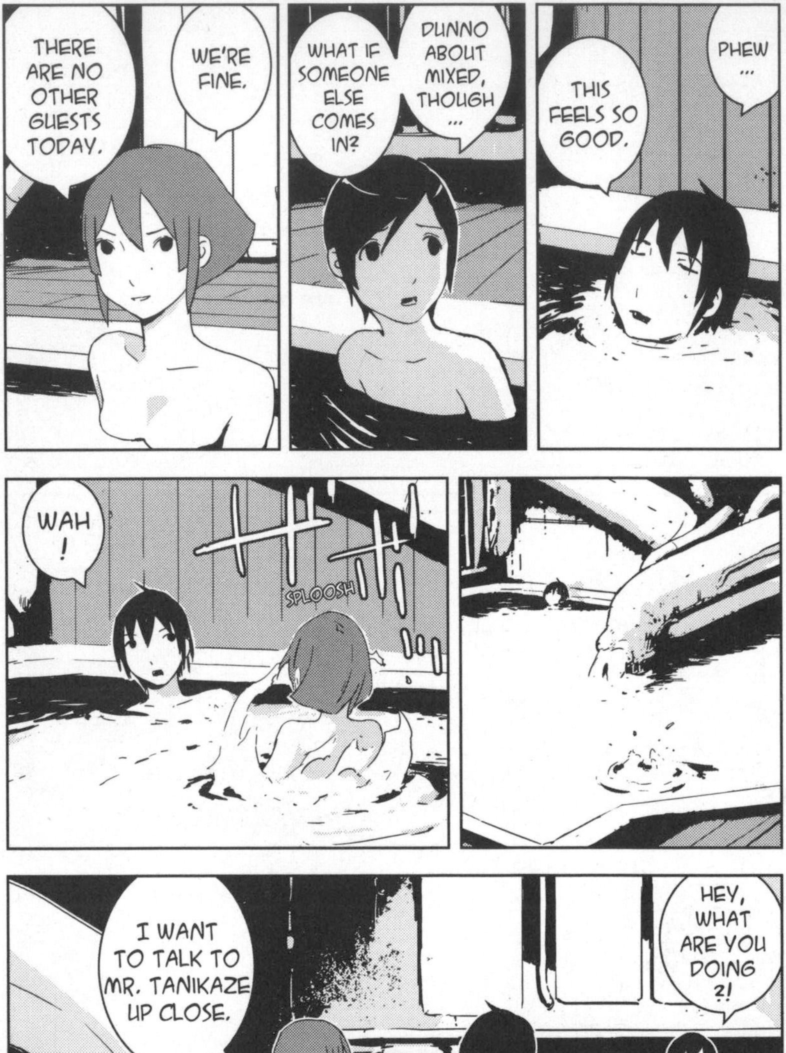
PHEW ...
THIS FEELS SO GOOD.
DUNNO ABOUT MIXED, THOUGH ...
WHAT IF SOMEONE ELSE COMES IN?
WE'RE FINE.
THERE ARE NO OTHER GUESTS TODAY.
WAH!
SPLOOSH
HEY, WHAT ARE YOU DOING?!
I WANT TO TALK TO MR. TANIKAZE UP CLOSE.

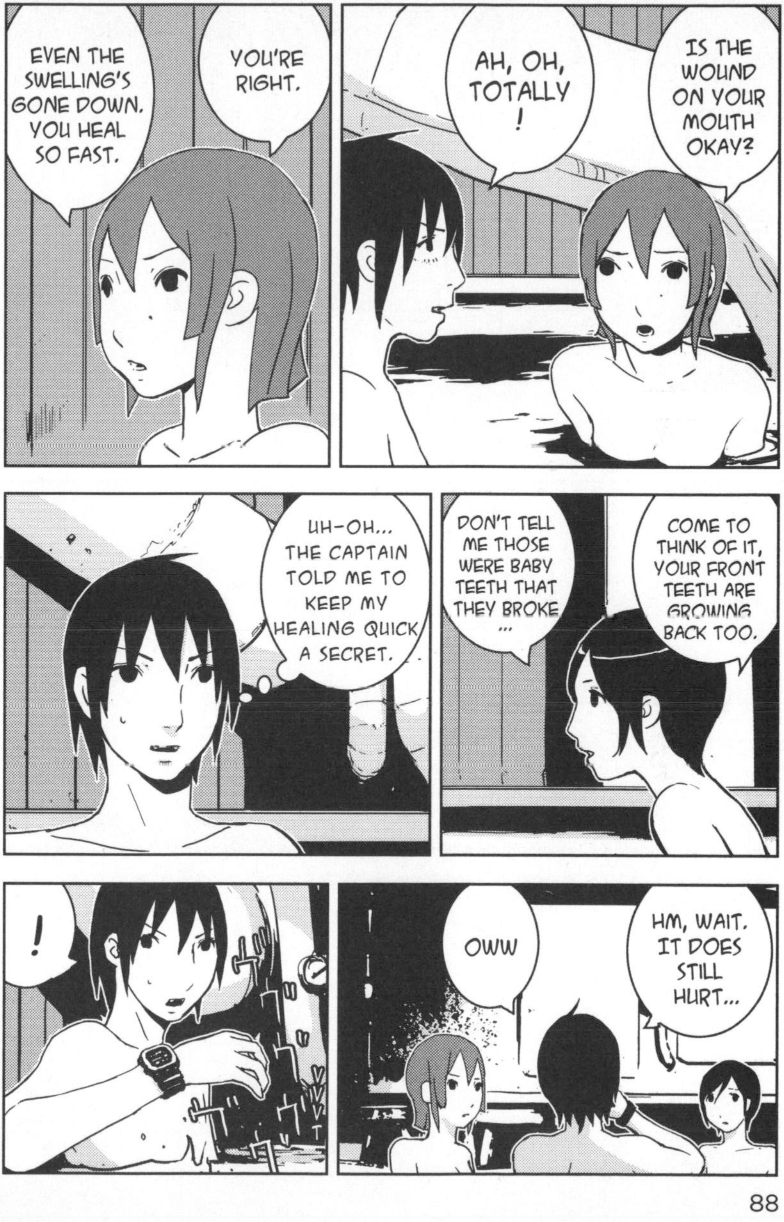
IS THE WOUND ON YOUR MOUTH OKAY?
AH, OH, TOTALLY!
YOU'RE RIGHT.
EVEN THE SWELLING'S GONE DOWN. YOU HEAL SO FAST.
COME TO THINK OF IT, YOUR FRONT TEETH ARE GROWING BACK TOO.
DON'T TELL ME THOSE WERE BABY TEETH THAT THEY BROKE...
UH-OH... THE CAPTAIN TOLD ME TO KEEP MY HEALING QUICK A SECRET.
HM, WAIT. IT DOES STILL HURT...
OWW
!

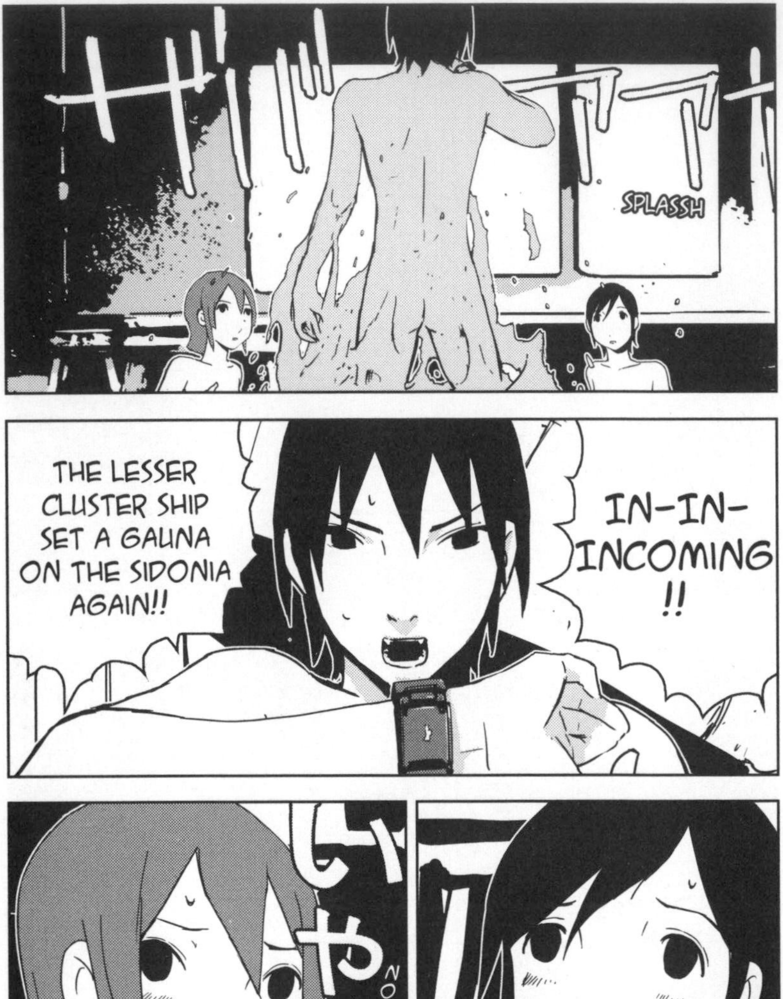
SPLASSH
IN-IN-INCOMING!!
THE LESSER CLUSTER SHIP SET A GAUNA ON THE SIDONIA AGAIN!!
N-NAGA-TE...
いやあああ
NOOOO

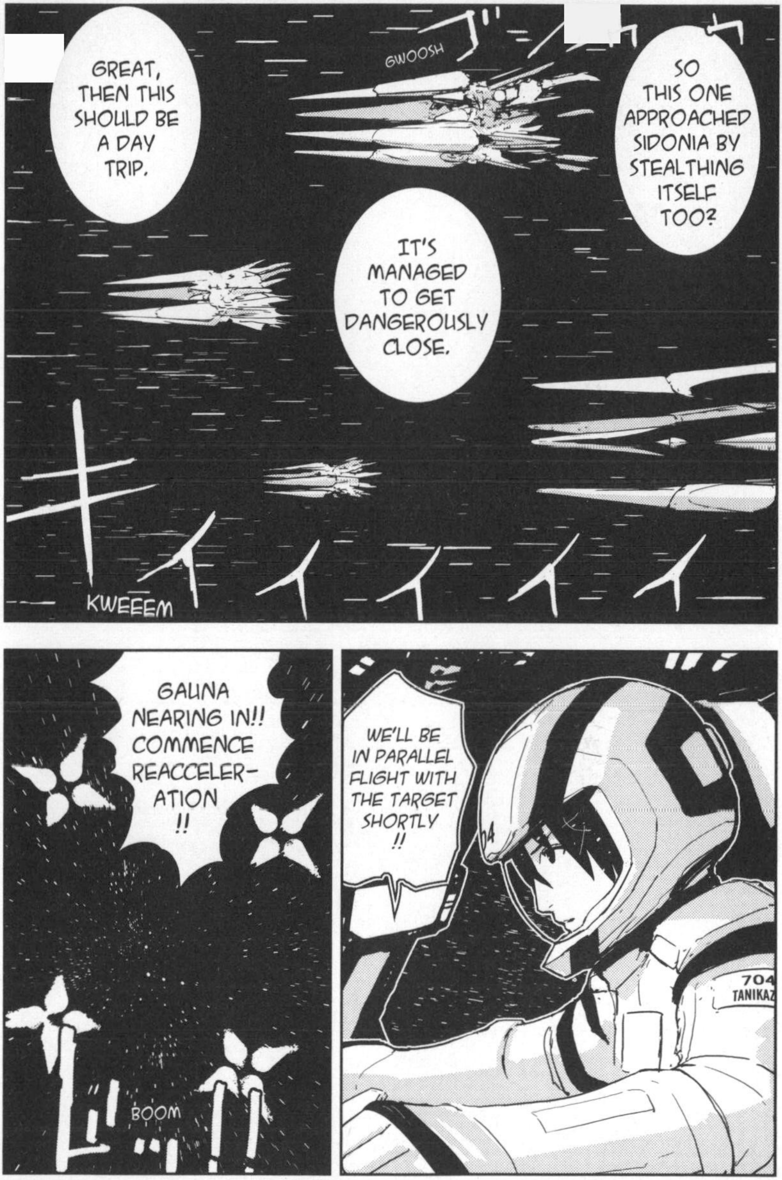
GWOOSH
SO THIS ONE APPROACHED SIDONIA BY STEALTHING ITSELF TOO?
IT'S MANAGED TO GET DANGEROUSLY CLOSE.
GREAT, THEN THIS SHOULD BE A DAY TRIP.
KWEEEM
WE'LL BE IN PARALLEL FLIGHT WITH THE TARGET SHORTLY !!
704 TANIKAZ
GAUNA NEARING IN!! COMMENCE REACCELER-ATION !!
BOOM

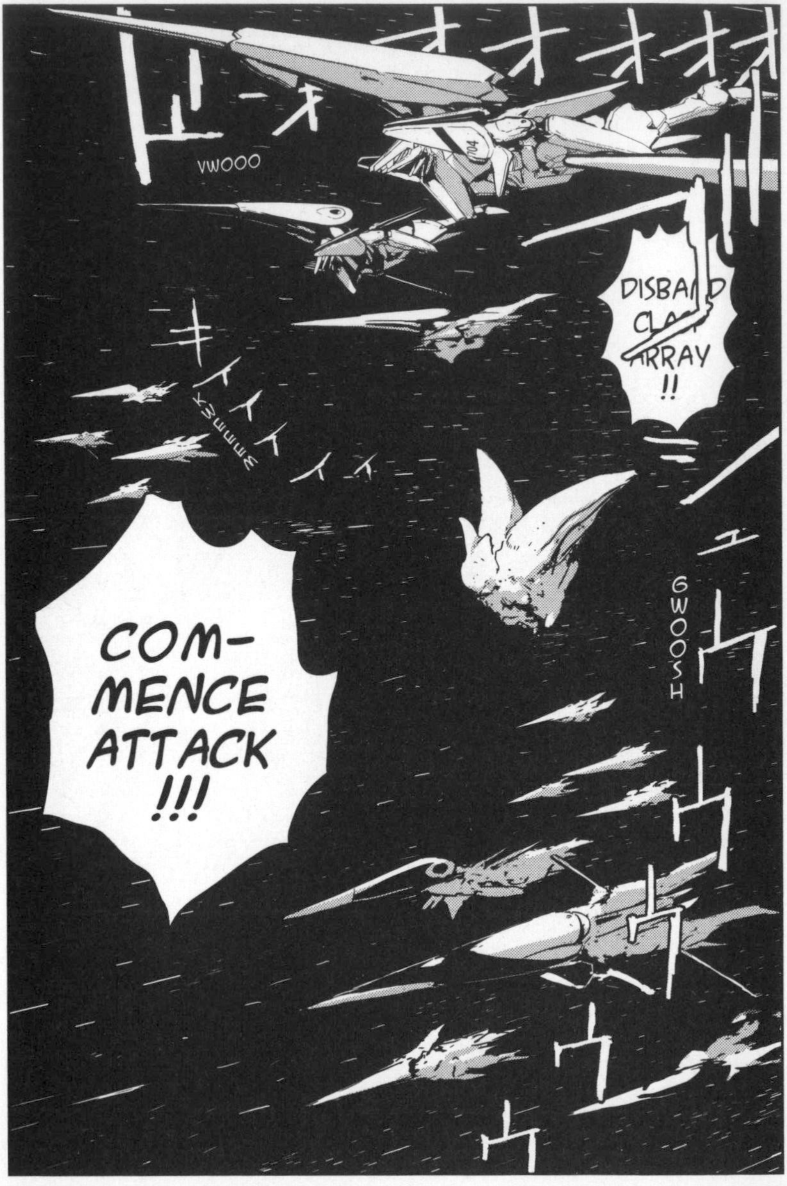
VWOOO
DISBAND CLA ARRAY !!
KSHHHH
GWOOSH
COM-MENCE ATTACK !!!

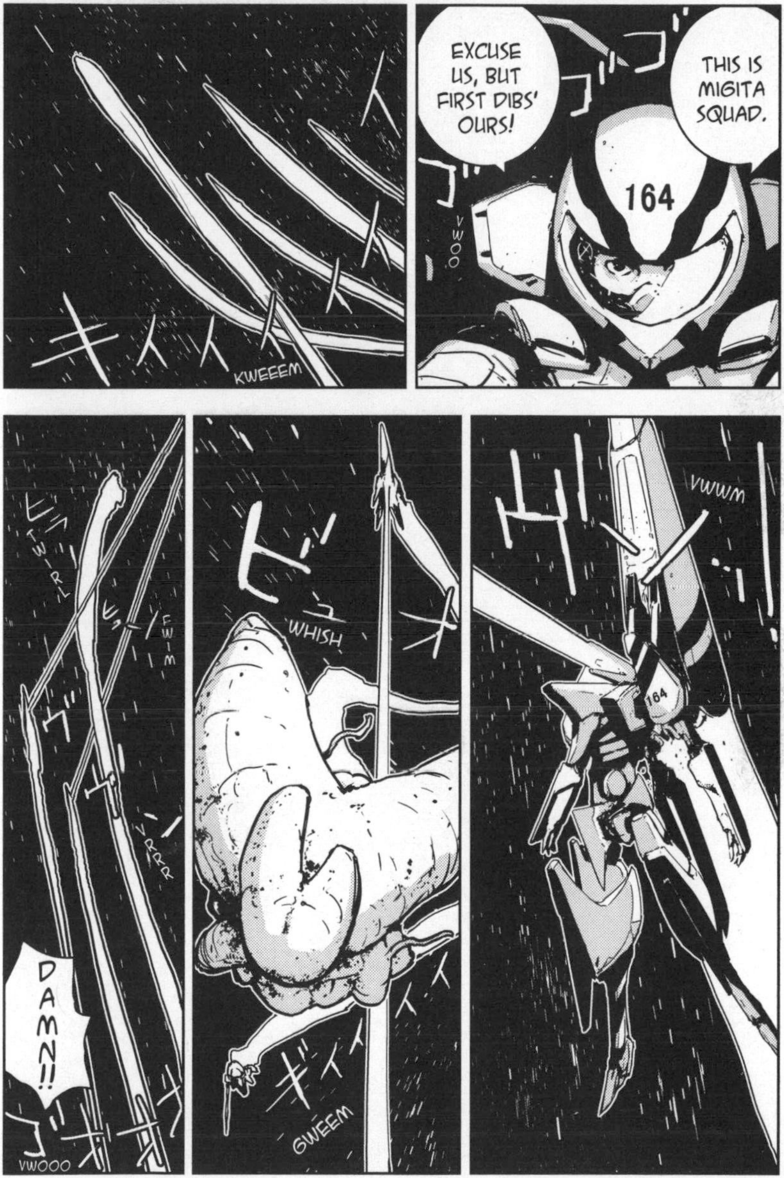
THIS IS MIGITA SQUAD.
EXCUSE US, BUT FIRST DIBS' OURS!
164
VWOO
KWEEEM
VWWM
164
WHISH
GWEEM
TWIRL
FWIM
VRRR
DAMN!!
VWOOO

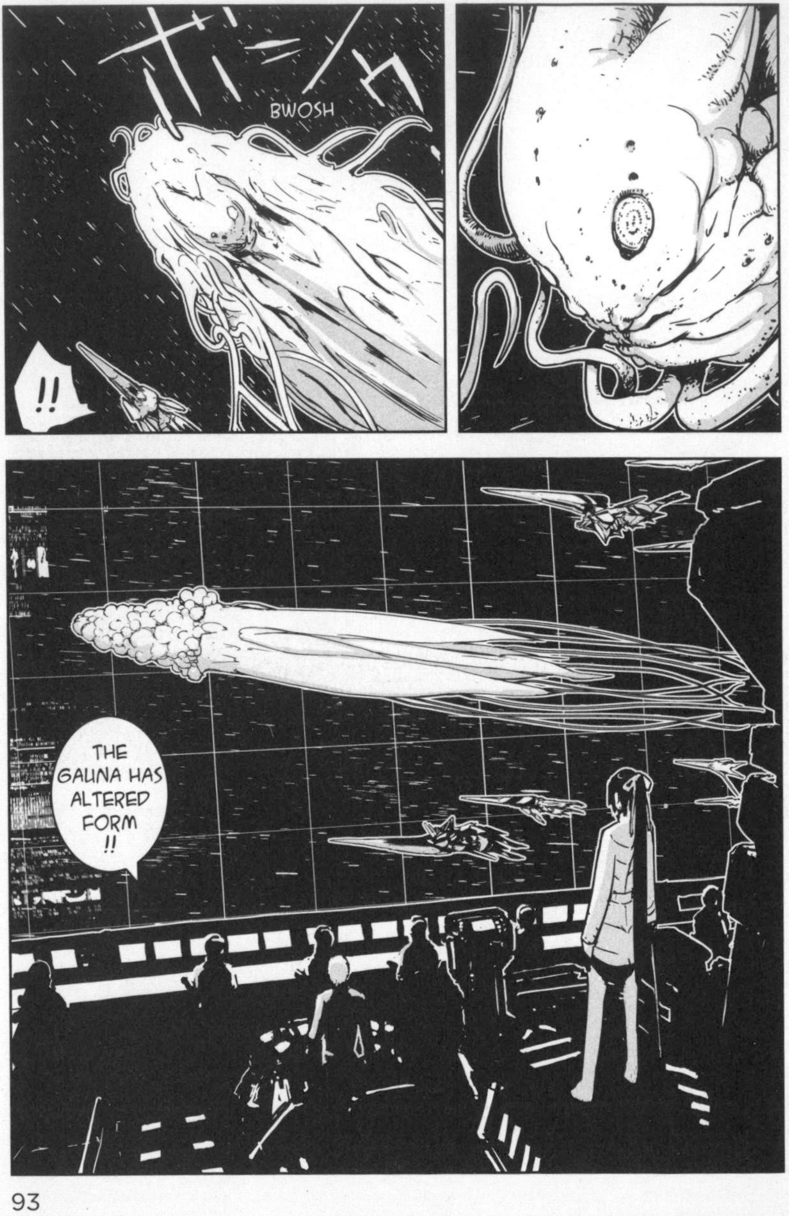
BWOSH
!!
THE GAUNA HAS ALTERED FORM !!

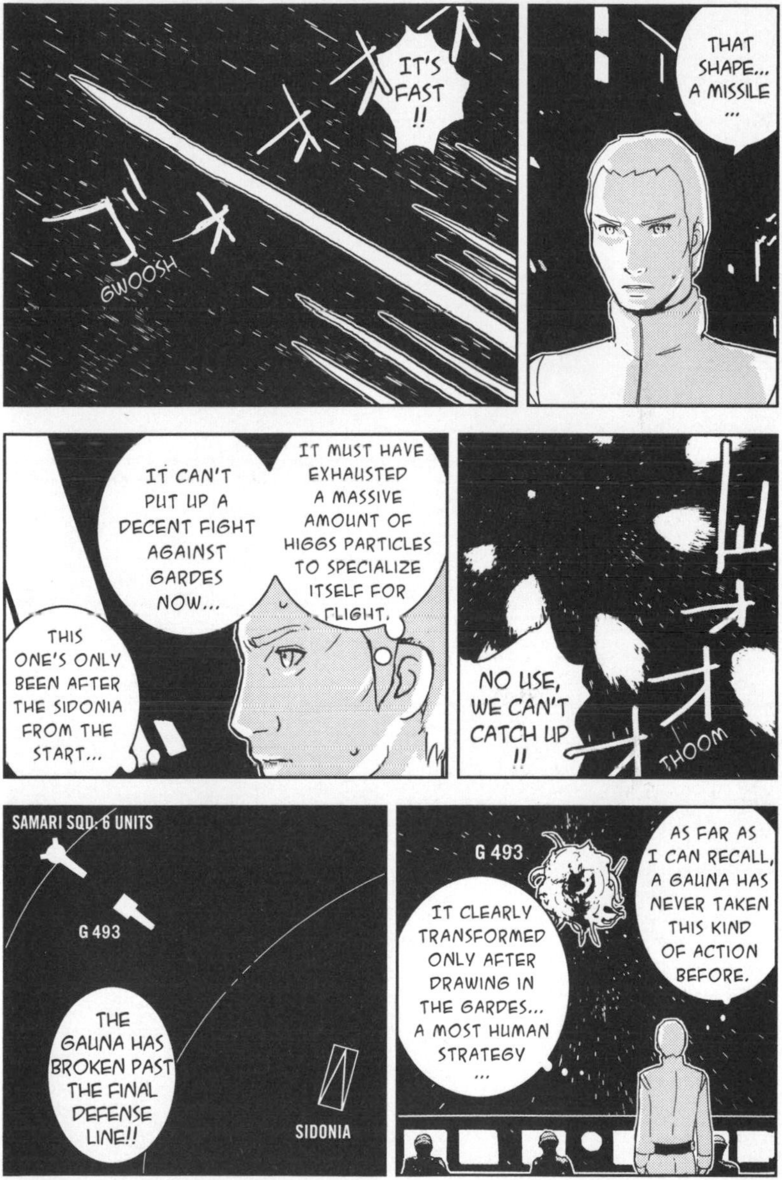
THAT SHAPE... A MISSILE ...
IT'S FAST !!
GWOOSH
NO USE, WE CAN'T CATCH UP !!
THOOM
IT MUST HAVE EXHAUSTED A MASSIVE AMOUNT OF HIGGS PARTICLES TO SPECIALIZE ITSELF FOR FLIGHT.
IT CAN'T PUT UP A DECENT FIGHT AGAINST GARDES NOW...
THIS ONE'S ONLY BEEN AFTER THE SIDONIA FROM THE START...
AS FAR AS I CAN RECALL, A GAUNA HAS NEVER TAKEN THIS KIND OF ACTION BEFORE.
G 493
IT CLEARLY TRANSFORMED ONLY AFTER DRAWING IN THE GARDES... A MOST HUMAN STRATEGY ...
SAMARI SQD: 6 UNITS
G 493
THE GAUNA HAS BROKEN PAST THE FINAL DEFENSE LINE!!
SIDONIA

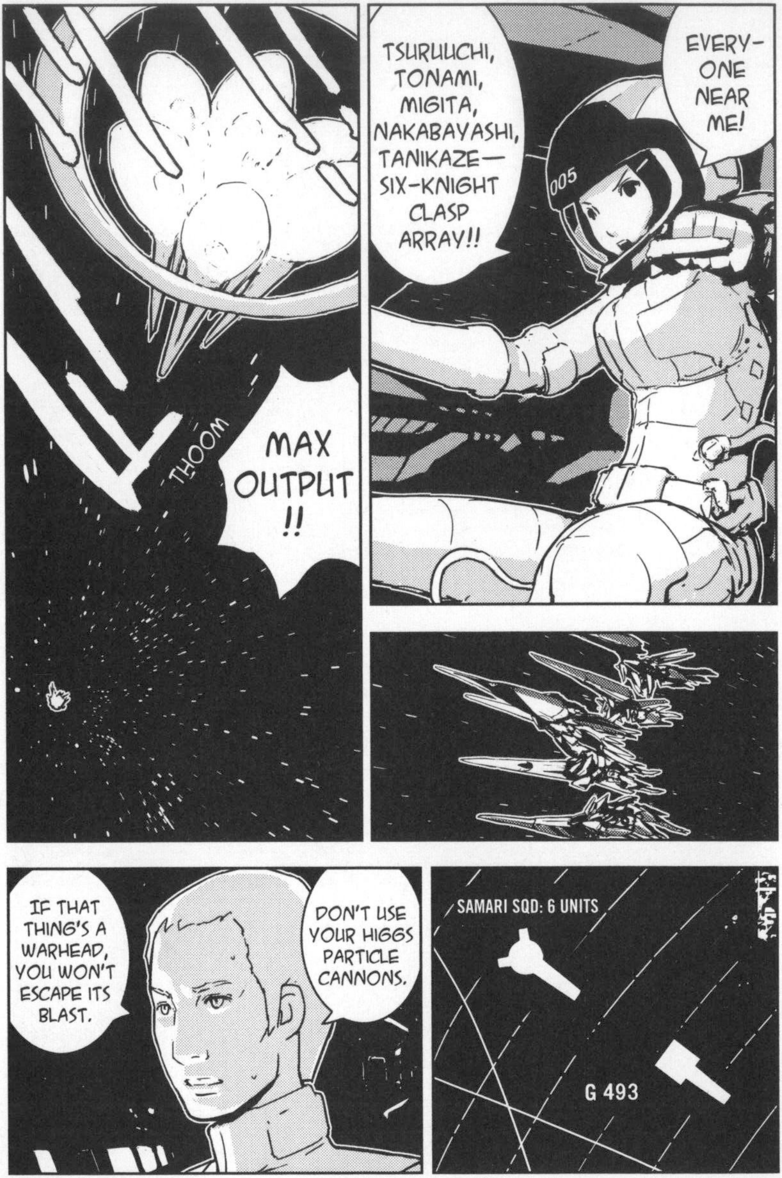
EVERY-ONE NEAR ME!
TSURUUCHI, TONAMI, MIGITA, NAKABAYASHI, TANIKAZE—SIX-KNIGHT CLASP ARRAY!!
THOOM
MAX OUTPUT !!
SAMARI SQD: 6 UNITS
G 493
DON'T USE YOUR HIGGS PARTICLE CANNONS.
IF THAT THING'S A WARHEAD, YOU WON'T ESCAPE ITS BLAST.

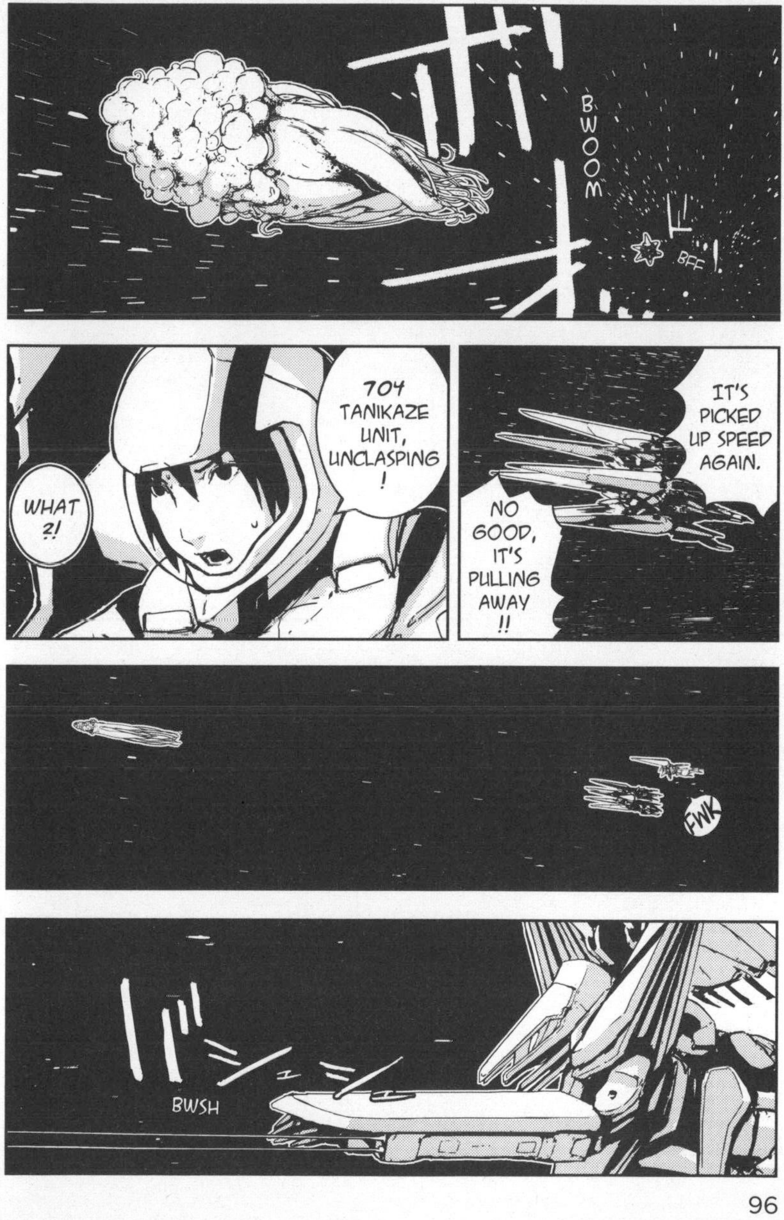
BWOOM
BFF
IT'S PICKED UP SPEED AGAIN.
NO GOOD, IT'S PULLING AWAY !!
704 TANIKAZE UNIT, UNCLASPING !
WHAT ?!
FWK
BWSH

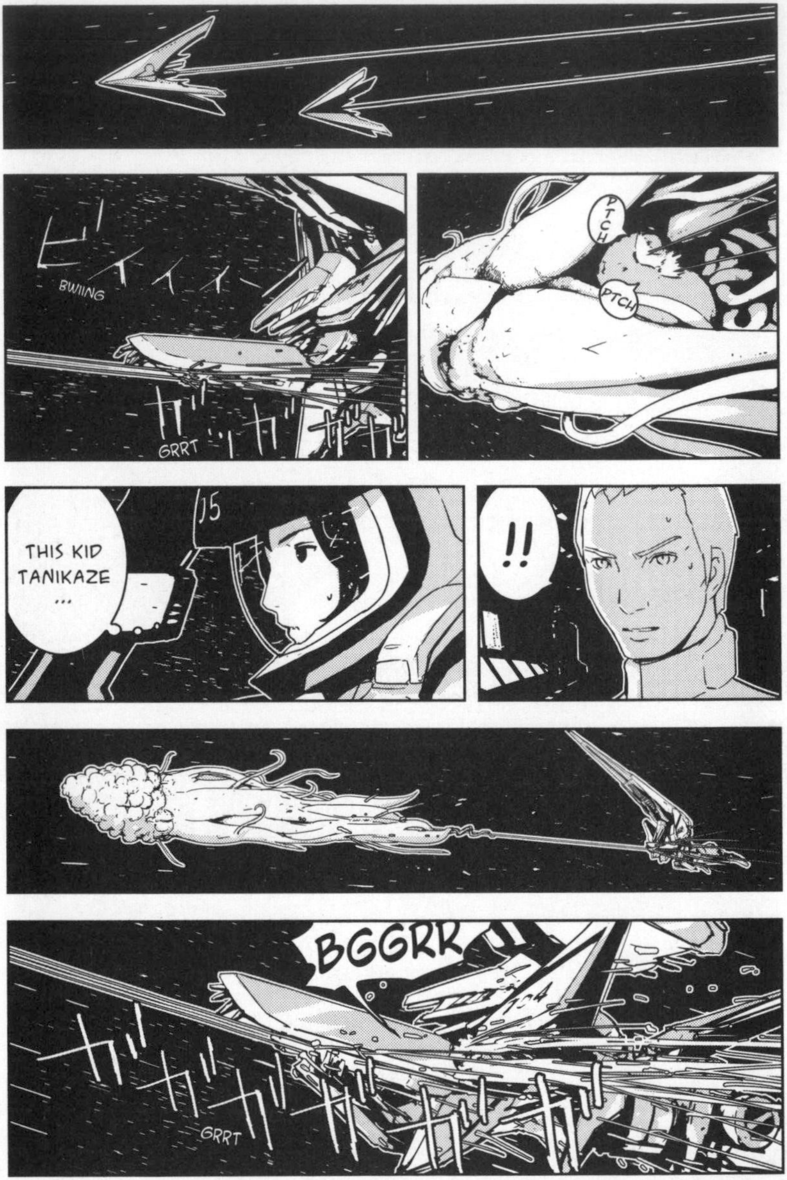
PTCH
PTCH
BWIING
GRRT
!!
THIS KID TANIKAZE ...
BGGRR
GRRT

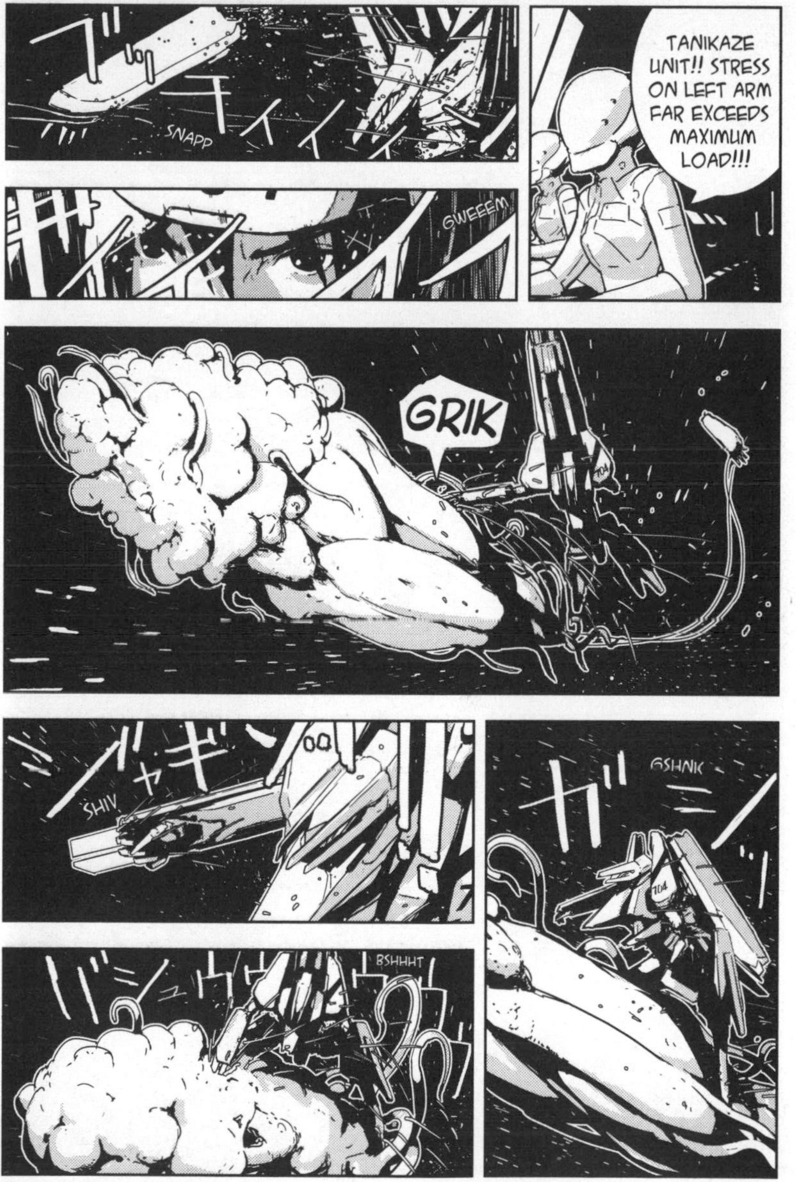
TANIKAZE UNIT!! STRESS ON LEFT ARM FAR EXCEEDS MAXIMUM LOAD!!!
SNAPP
GWEEEM
GRIK
GSHNK
SHIV
BSHHHT

GWEEEM
BLUPP
ONE OUT !!
GET AWAY !!!
KRIK
KABOOOM
ROAR

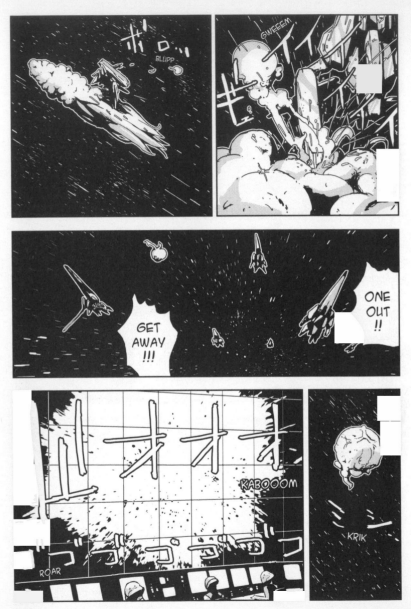

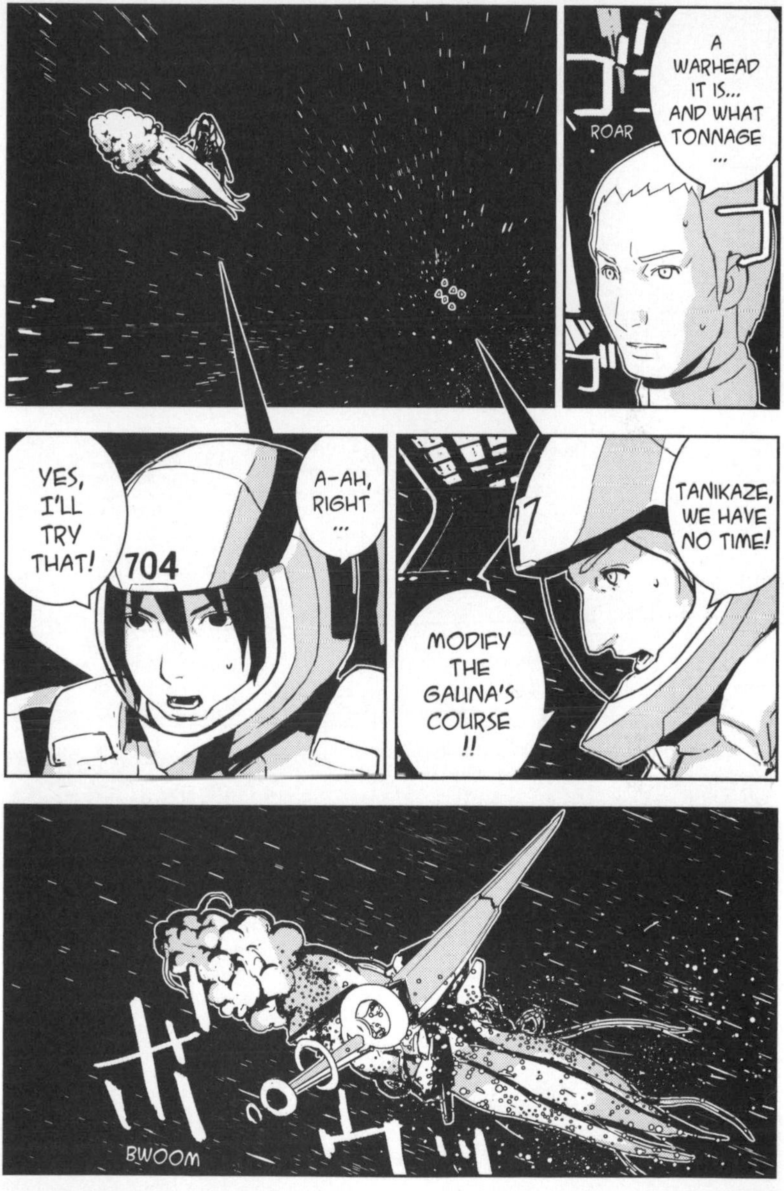
A WARHEAD IT IS... AND WHAT TONNAGE ...
ROAR
TANIKAZE, WE HAVE NO TIME!
MODIFY THE GAUNA'S COURSE !!
A-AH, RIGHT ...
YES, I'LL TRY THAT!
704
BWOOM

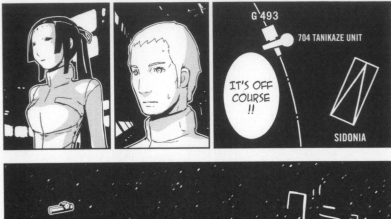

G 493

704 TANIKAZE UNIT

IT'S OFF COURSE!!

SIDONIA

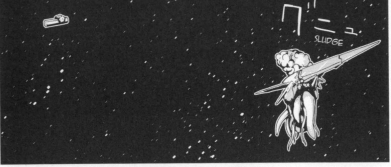

SLUDGE

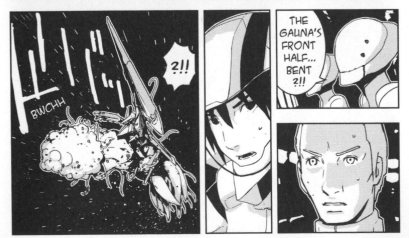

BWCHH

?!!

THE GAUNA'S FRONT HALF... BENT?!!

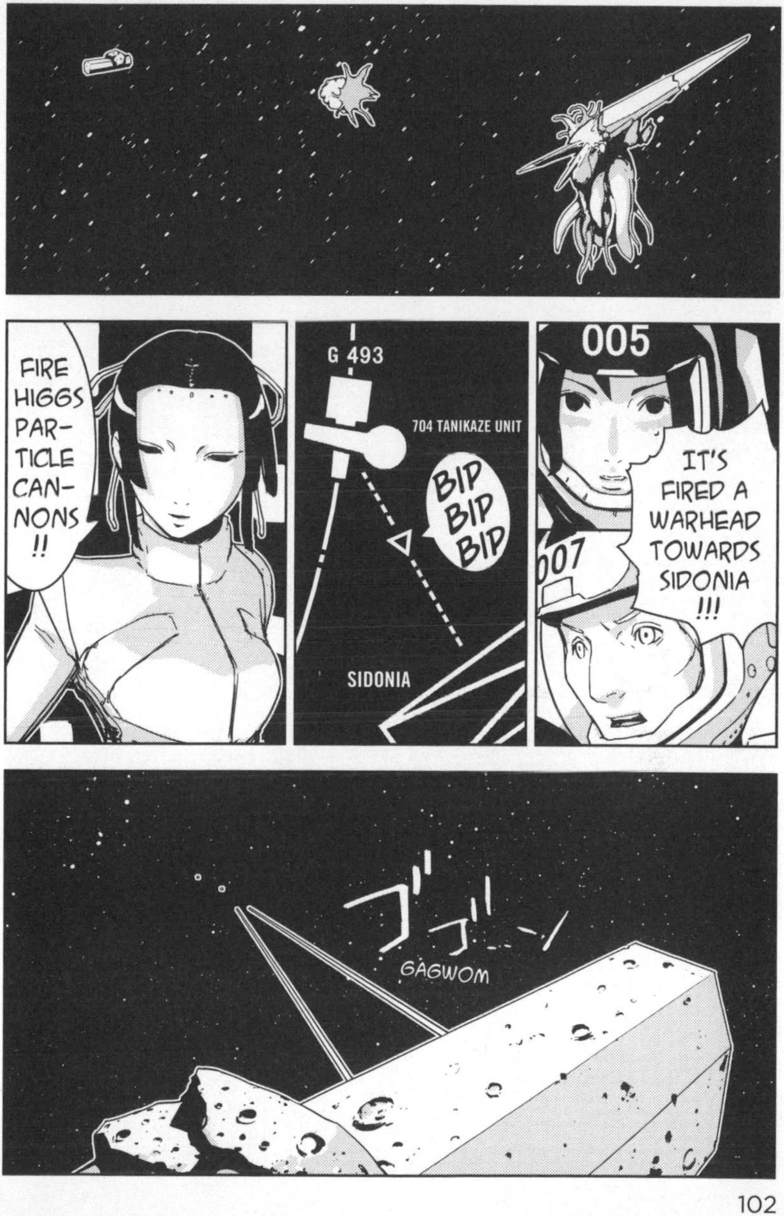
005
IT'S FIRED A WARHEAD TOWARDS SIDONIA !!!
007
G 493
704 TANIKAZE UNIT
BIP BIP BIP
SIDONIA
FIRE HIGGS PARTICLE CANNONS !!
GAGWOM

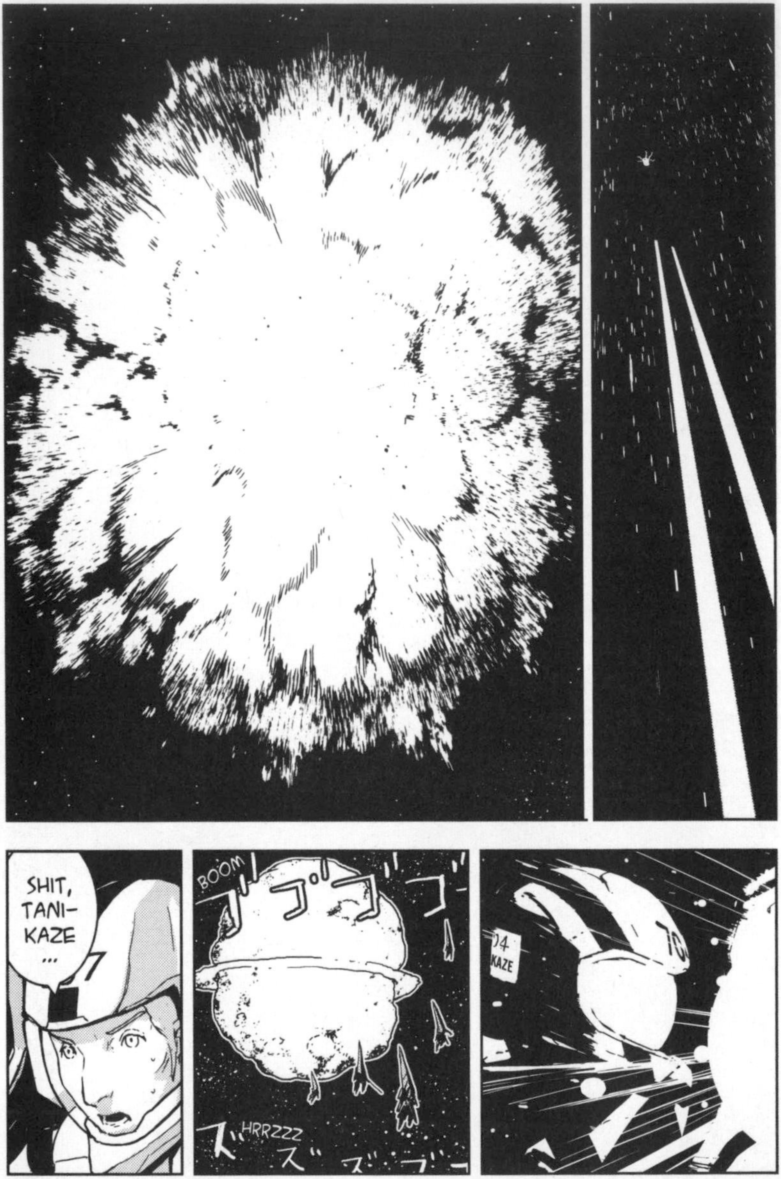
KAZE
BOOM
ゴゴゴー
HRRZZZ
ズズズー
SHIT, TANI-KAZE ...

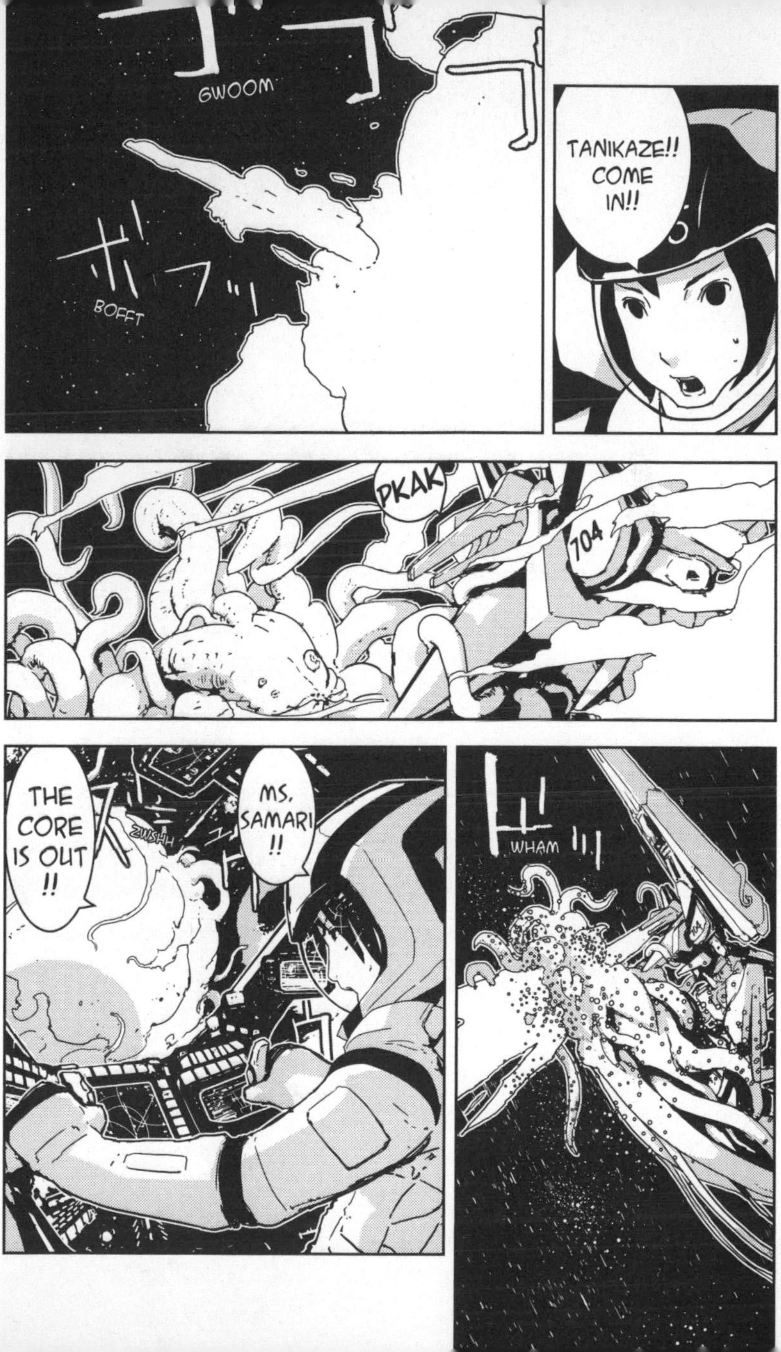
TANIKAZE!! COME IN!!
GWOOM
BOFFT
PKAK
704
WHAM
MS. SAMARI !!
ZWSHH
THE CORE IS OUT !!

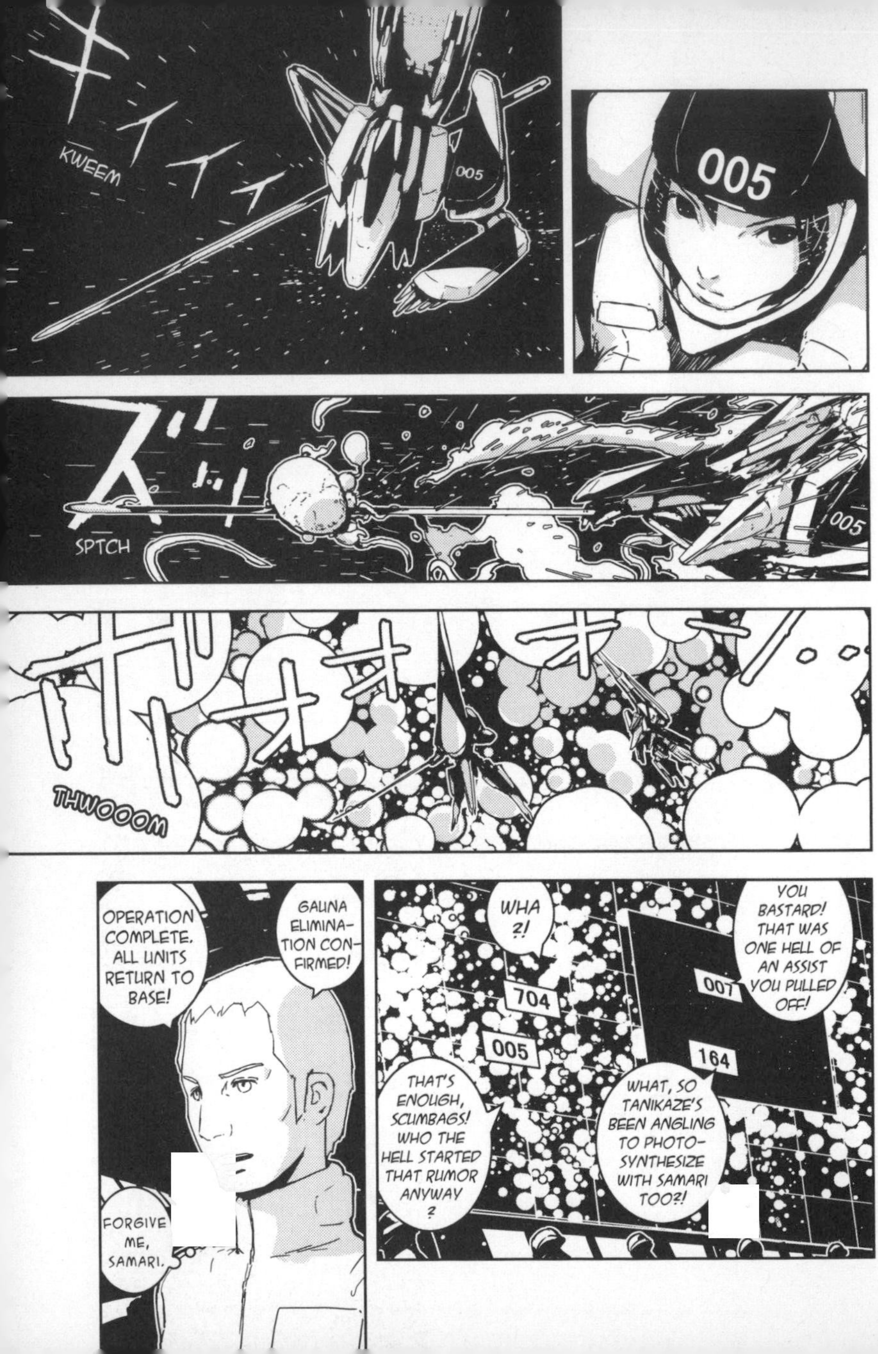

005
KWEEM
005
SPTCH
005
THWOOOM
YOU BASTARD! THAT WAS ONE HELL OF AN ASSIST YOU PULLED OFF!
WHA ?!
007
704
164
005
WHAT, SO TANIKAZE'S BEEN ANGLING TO PHOTO-SYNTHESIZE WITH SAMARI TOO?!
THAT'S ENOUGH, SCUMBAGS! WHO THE HELL STARTED THAT RUMOR ANYWAY ?
GAUNA ELIMINATION CONFIRMED!
OPERATION COMPLETE. ALL UNITS RETURN TO BASE!
FORGIVE ME, SAMARI.

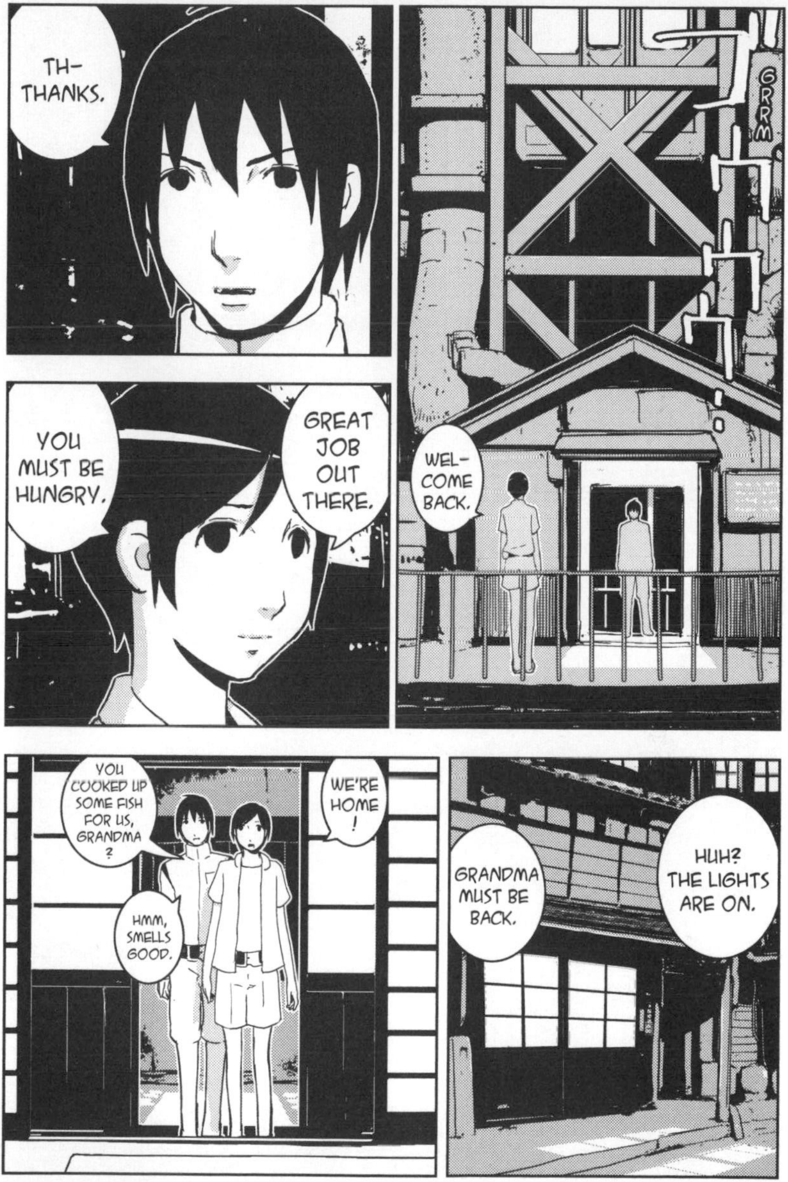
GRRM
WEL-COME BACK.
TH- THANKS.
GREAT JOB OUT THERE.
YOU MUST BE HUNGRY.
HUH? THE LIGHTS ARE ON.
GRANDMA MUST BE BACK.
WE'RE HOME!
YOU COOKED UP SOME FISH FOR US, GRANDMA?
HMM, SMELLS GOOD.

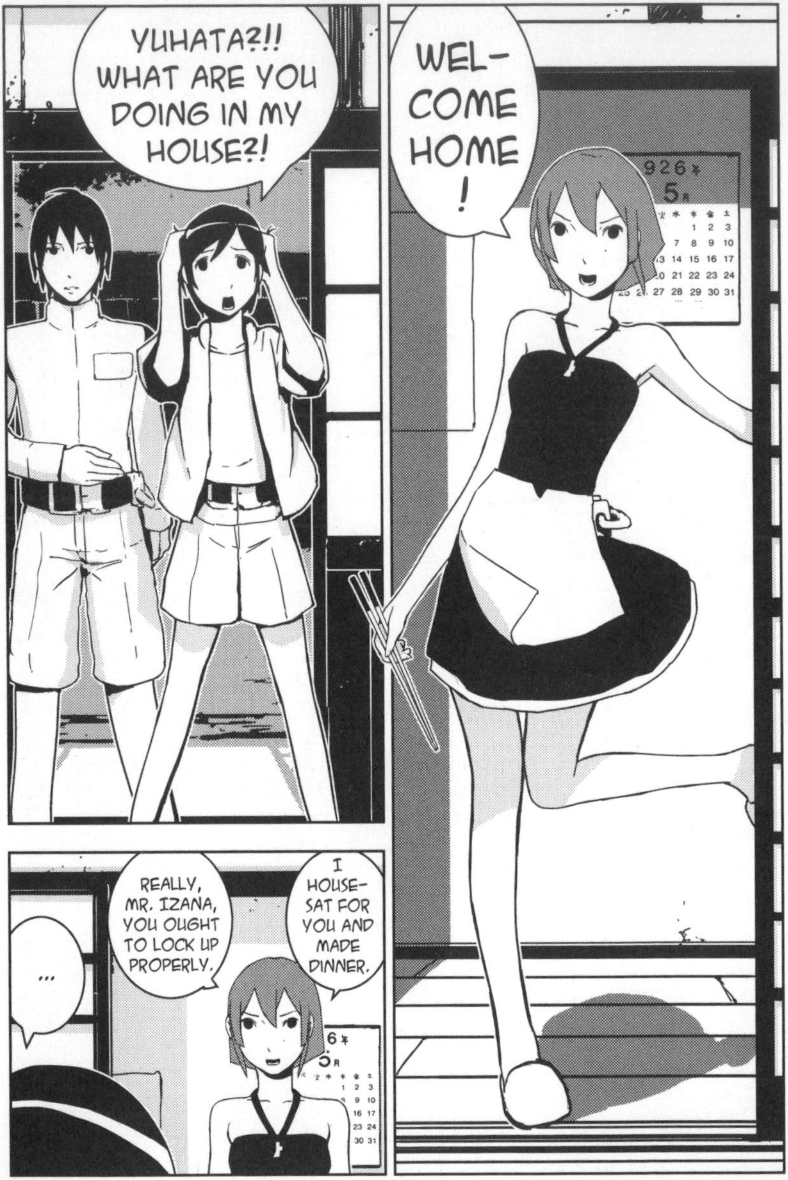
WEL-COME HOME!
YUHATA?!! WHAT ARE YOU DOING IN MY HOUSE?!
I HOUSE-SAT FOR YOU AND MADE DINNER.
REALLY, MR. IZANA, YOU OUGHT TO LOCK UP PROPERLY.
...

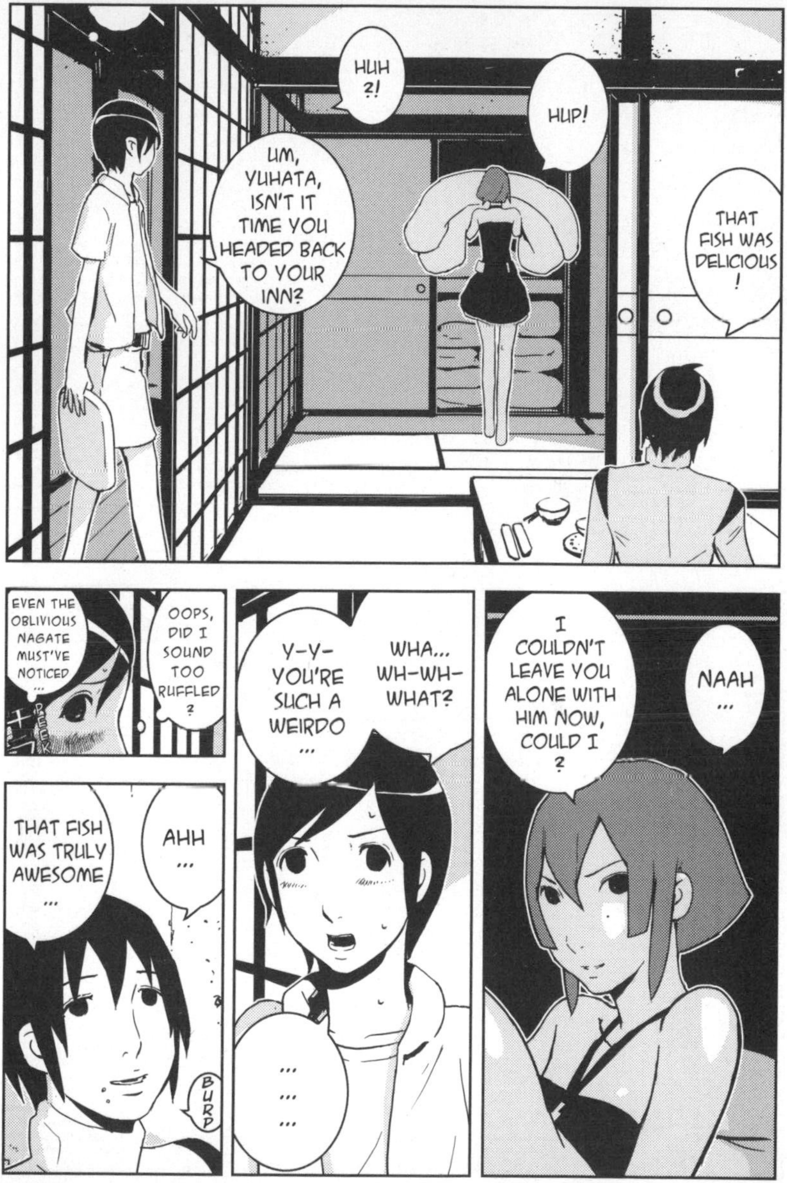

THAT FISH WAS DELICIOUS!
HUP!
UM, YUHATA, ISN'T IT TIME YOU HEADED BACK TO YOUR INN?
HUH?!
NAAH...
I COULDN'T LEAVE YOU ALONE WITH HIM NOW, COULD I?
WHA... WH-WH-WHAT?
Y-Y-YOU'RE SUCH A WEIRDO...
... ...
OOPS, DID I SOUND TOO RUFFLED?
EVEN THE OBLIVIOUS NAGATE MUST'VE NOTICED...
PEEK
AHH...
THAT FISH WAS TRULY AWESOME...
BURP

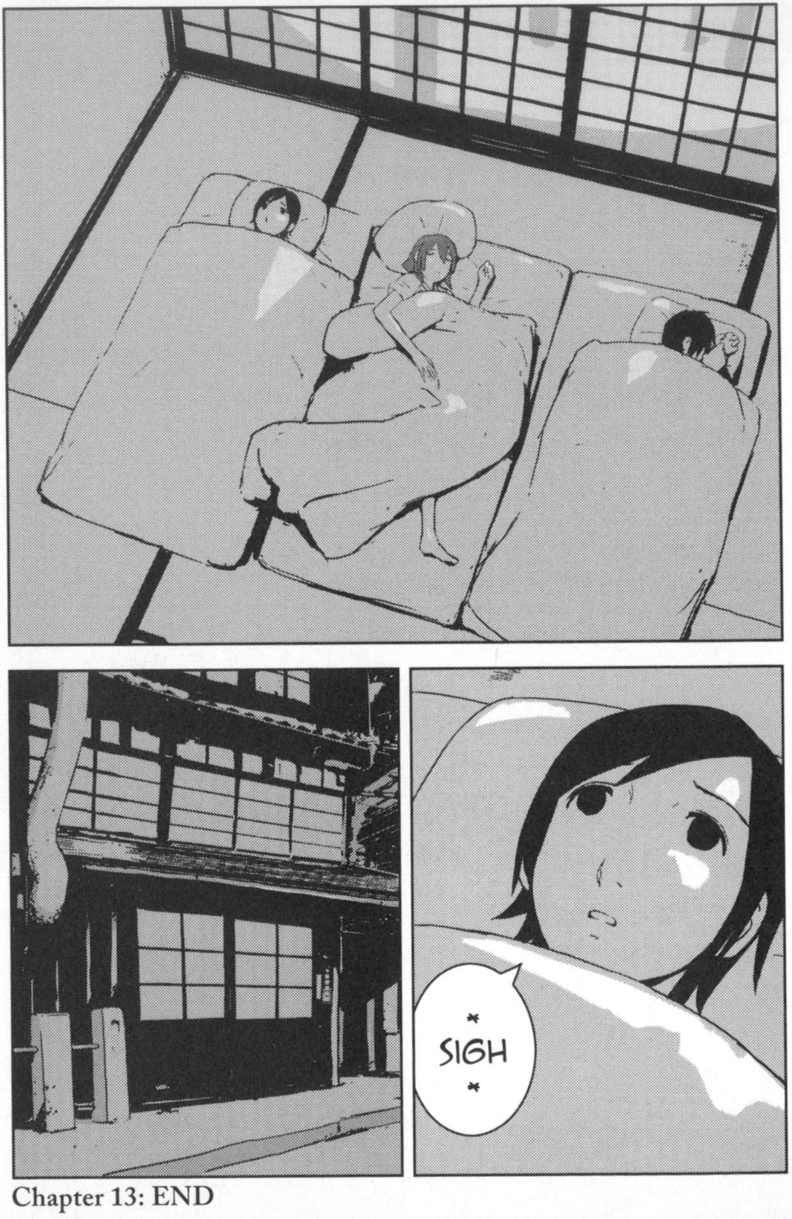
SIGH

Chapter 13: END

シドニアの騎士
KNIGHTS OF SIDONIA

Chapter 14: The Ship's Immortal Crew

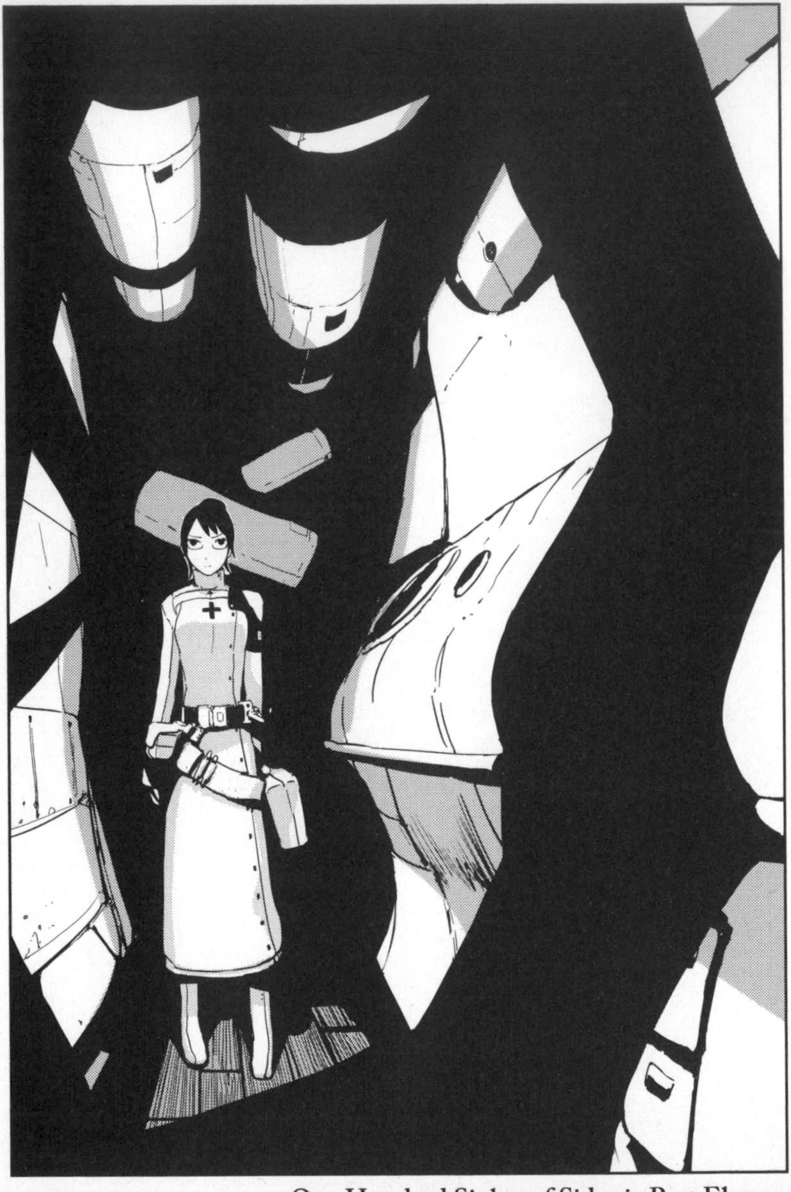

One Hundred Sights of Sidonia Part Eleven:
Pilot Dormitories Hidden Passage

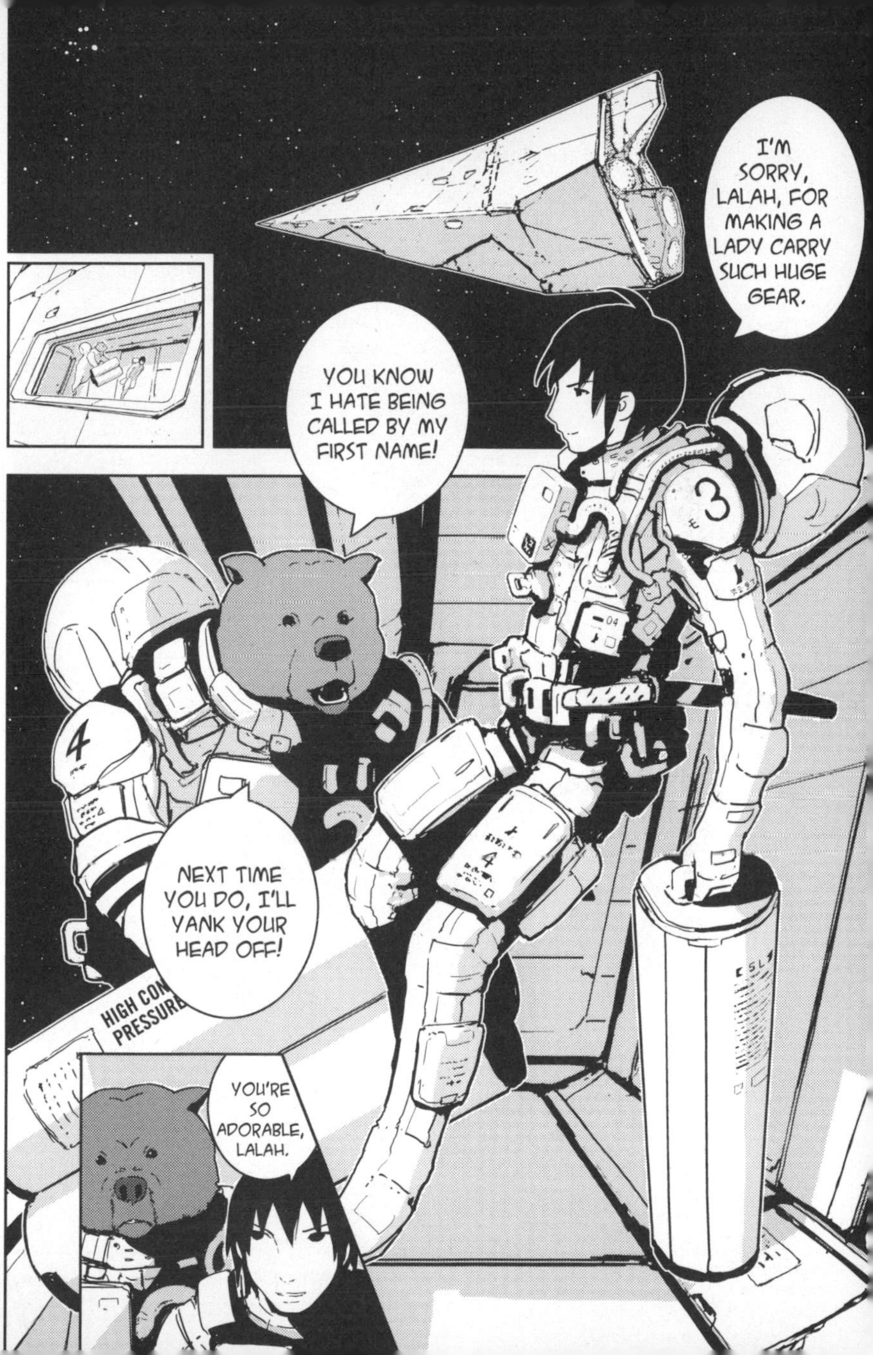
I'M SORRY, LALAH, FOR MAKING A LADY CARRY SUCH HUGE GEAR.
YOU KNOW I HATE BEING CALLED BY MY FIRST NAME!
NEXT TIME YOU DO, I'LL YANK YOUR HEAD OFF!
YOU'RE SO ADORABLE, LALAH.

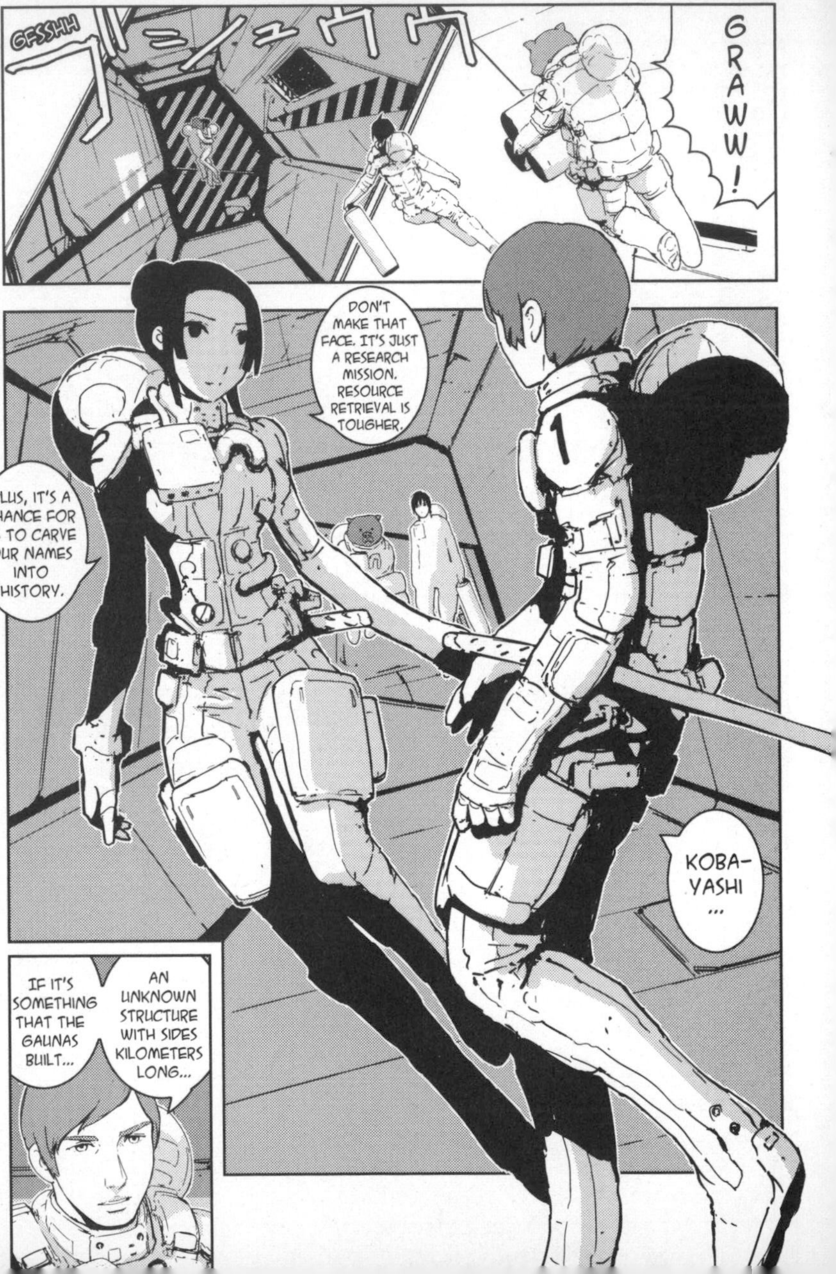
GRAWW!
GFSSHH
DON'T MAKE THAT FACE. IT'S JUST A RESEARCH MISSION. RESOURCE RETRIEVAL IS TOUGHER.
PLUS, IT'S A CHANCE FOR US TO CARVE OUR NAMES INTO HISTORY.
KOBA-YASHI ...
AN UNKNOWN STRUCTURE WITH SIDES KILOMETERS LONG...
IF IT'S SOMETHING THAT THE GAUNAS BUILT...

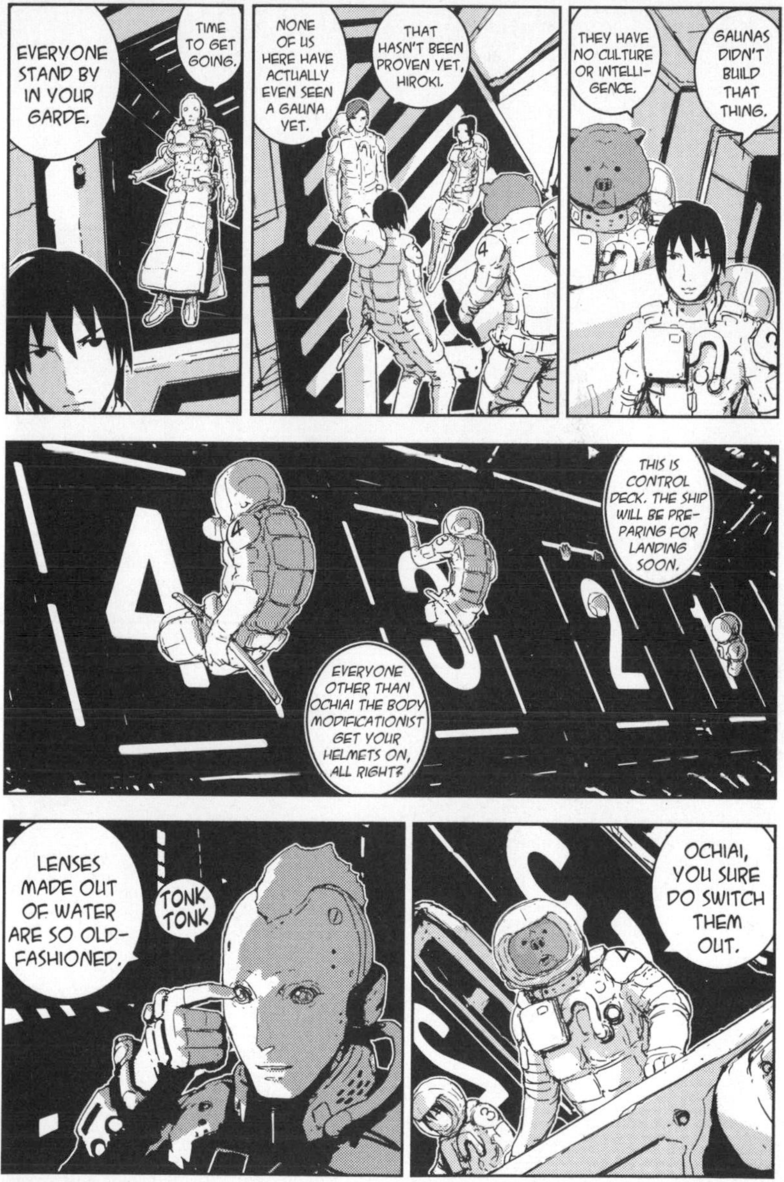
GAUNAS DIDN'T BUILD THAT THING.
THEY HAVE NO CULTURE OR INTELLIGENCE.
THAT HASN'T BEEN PROVEN YET, HIROKI.
NONE OF US HERE HAVE ACTUALLY EVEN SEEN A GAUNA YET.
TIME TO GET GOING.
EVERYONE STAND BY IN YOUR GARDE.
THIS IS CONTROL DECK. THE SHIP WILL BE PREPARING FOR LANDING SOON.
EVERYONE OTHER THAN OCHIAI THE BODY MODIFICATIONIST GET YOUR HELMETS ON, ALL RIGHT?
OCHIAI, YOU SURE DO SWITCH THEM OUT.
TONK TONK
LENSES MADE OUT OF WATER ARE SO OLD-FASHIONED.

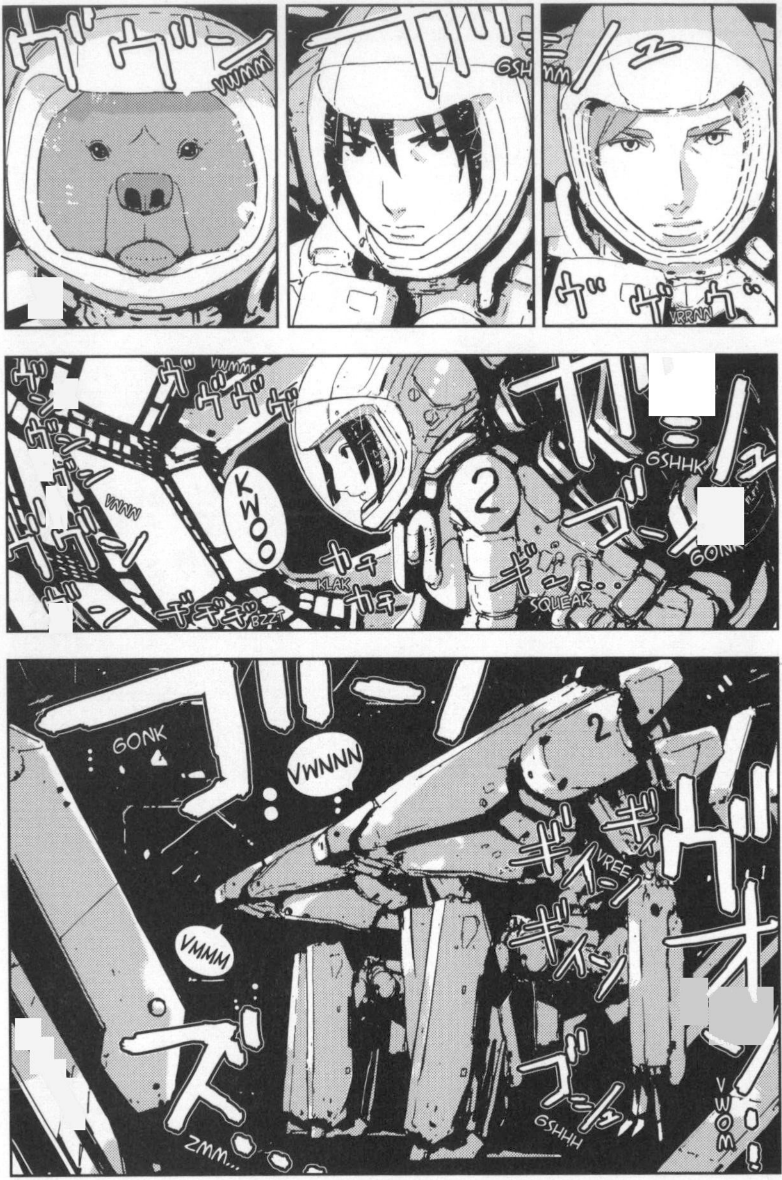
VRRNN
GSHMM
VWMM
GSHHK
GONK
SQUEAK
KWOO
KLAK
BZZT
VWMM
VNNN
VWOM!
GSHHH
VREE
VWNNN
GONK
VMMM
ZMM...

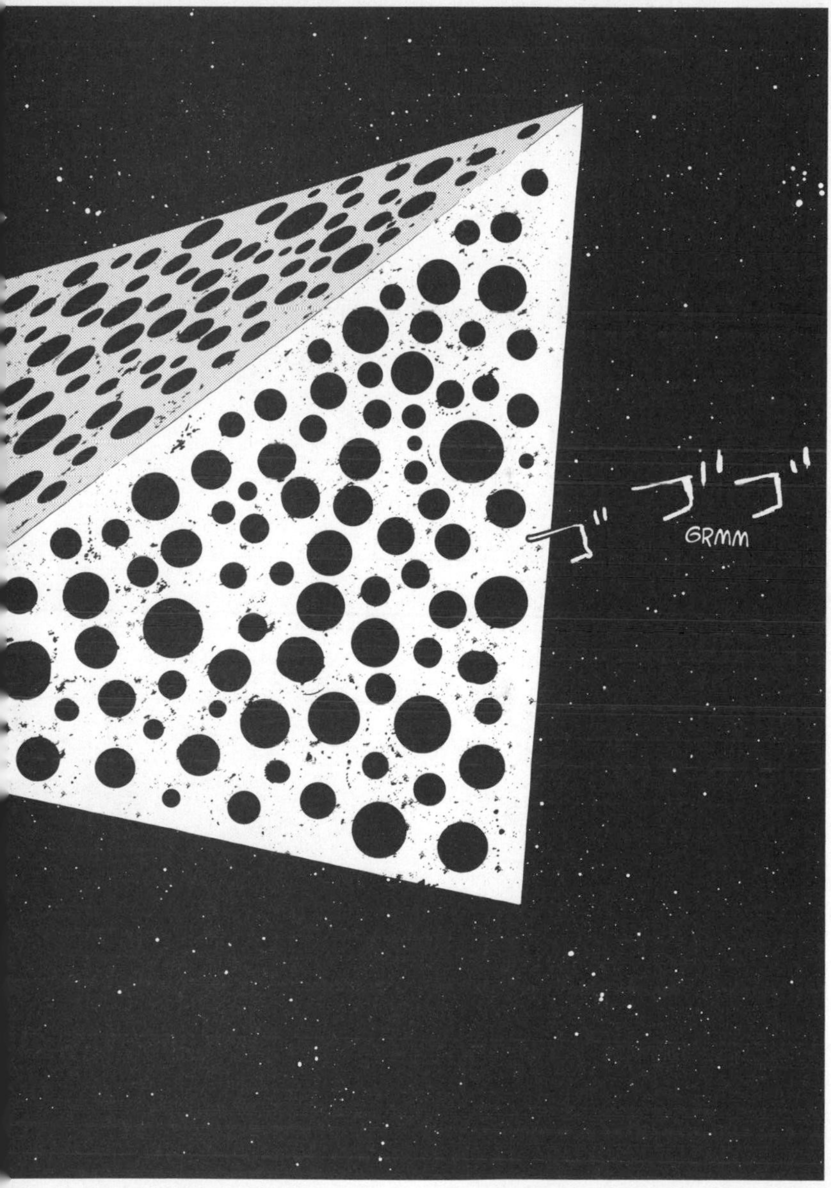
ゴゴゴ
GRMM

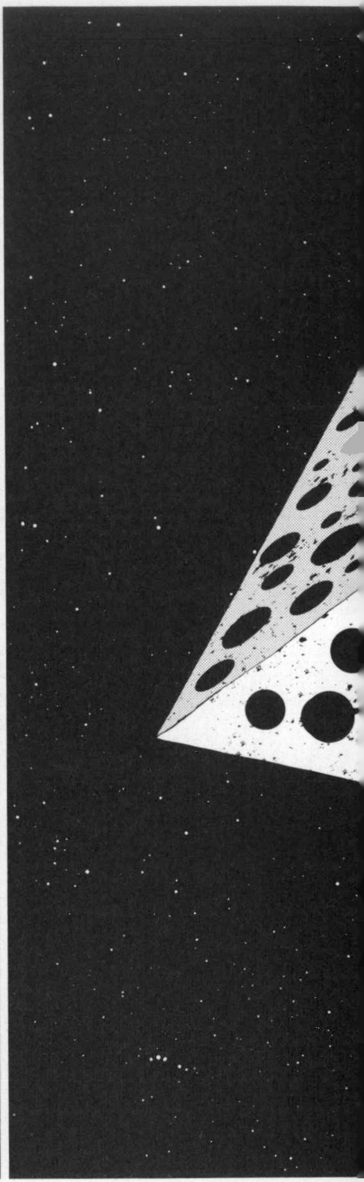
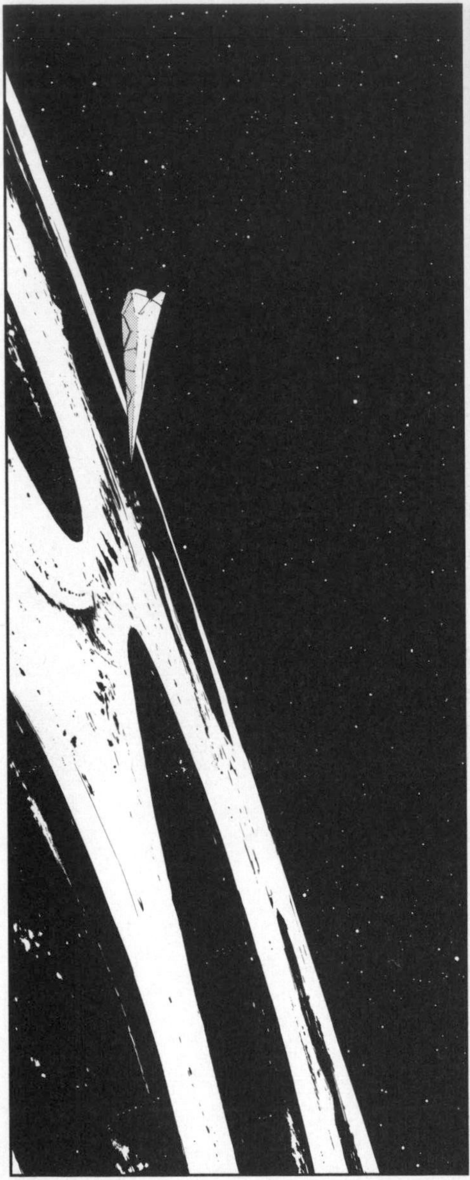

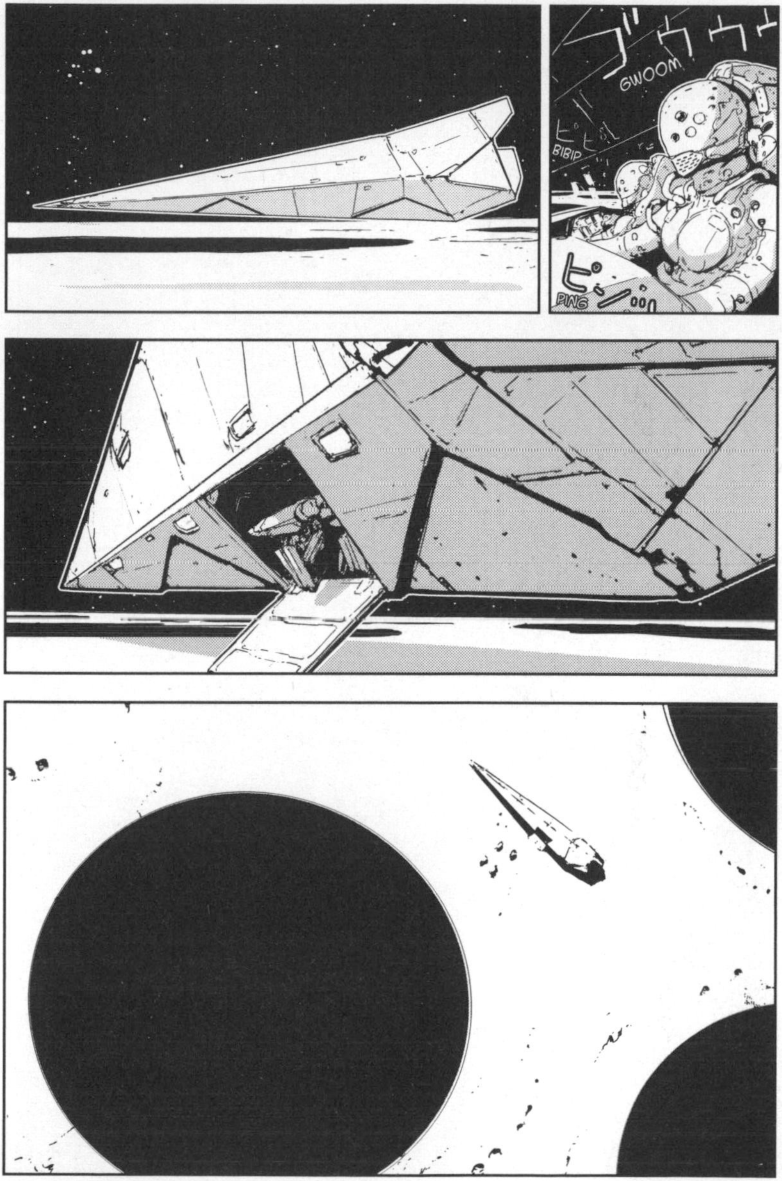
ゴウウ
GWOOM
ピピ
BIBIP
ピーン
PING

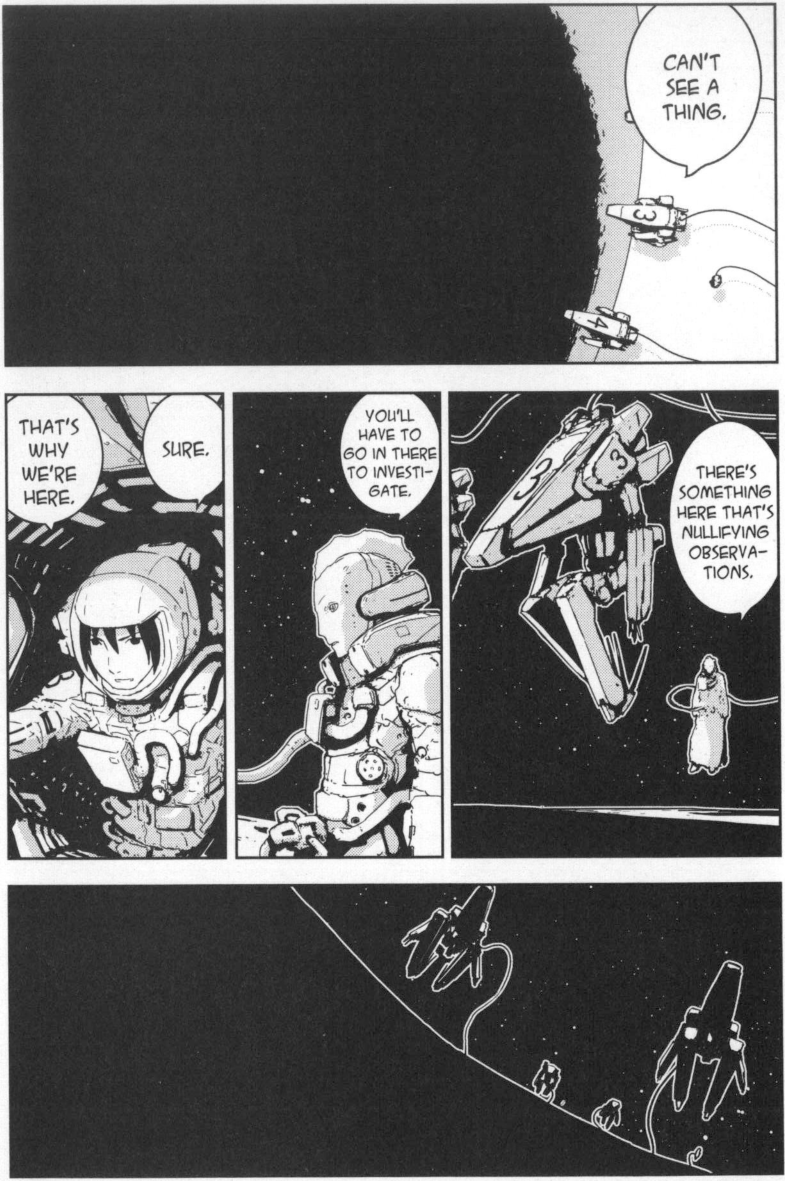
CAN'T SEE A THING.
THERE'S SOMETHING HERE THAT'S NULLIFYING OBSERVA-TIONS.
YOU'LL HAVE TO GO IN THERE TO INVESTI-GATE.
SURE.
THAT'S WHY WE'RE HERE.

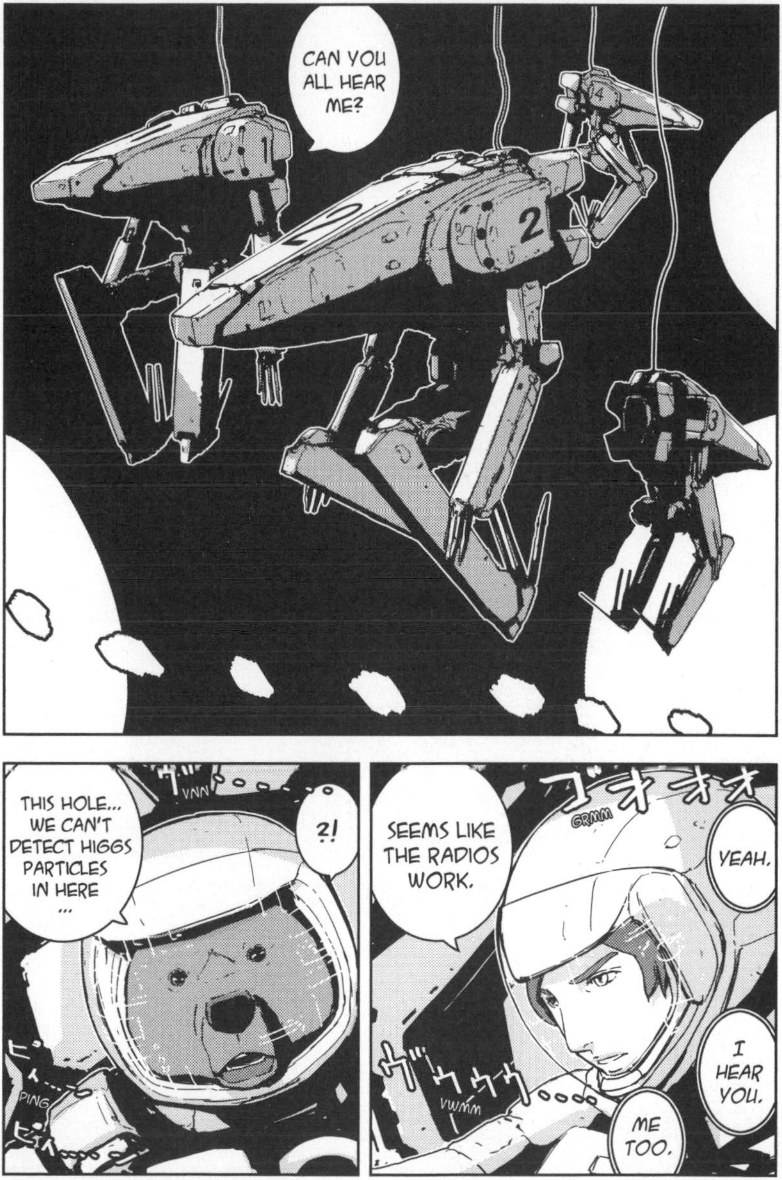
CAN YOU ALL HEAR ME?
ゴオオオ
GRMM
YEAH.
SEEMS LIKE THE RADIOS WORK.
I HEAR YOU.
ME TOO.
ヴゥゥゥ
VWMM
?!
VNN
THIS HOLE... WE CAN'T DETECT HIGGS PARTICLES IN HERE ...
PING

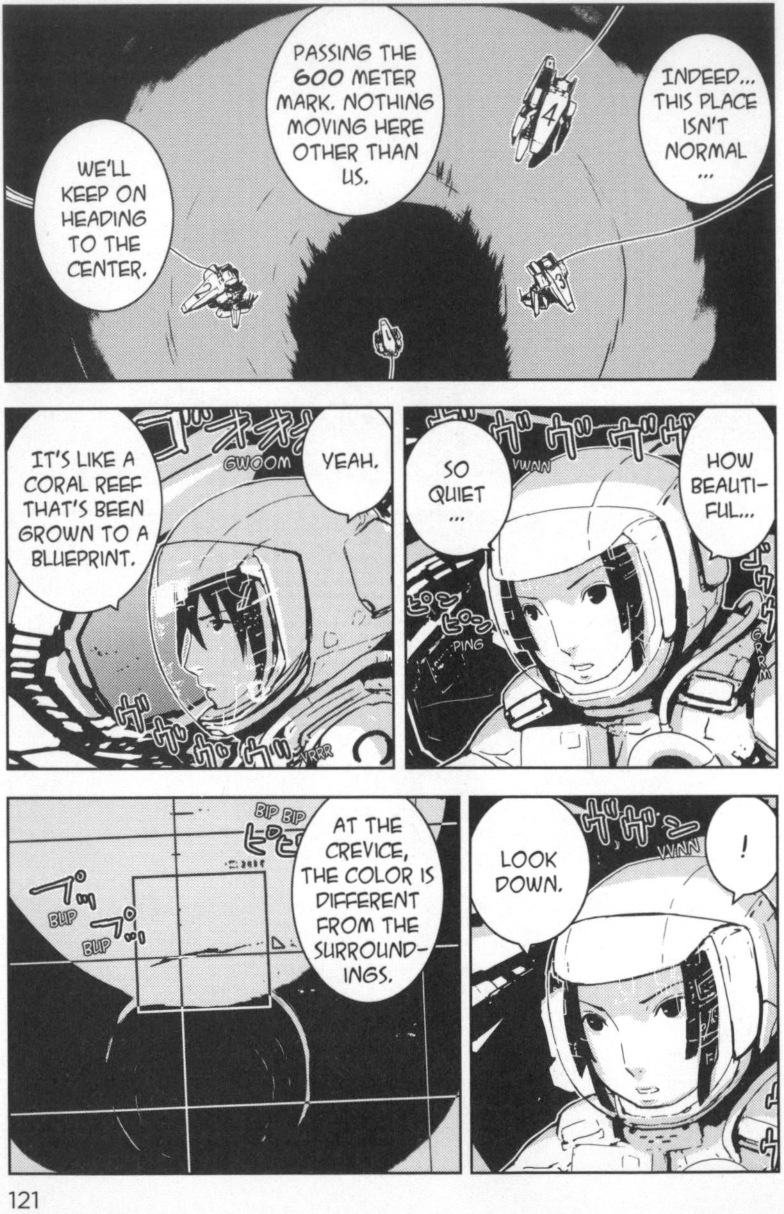
INDEED... THIS PLACE ISN'T NORMAL ...
PASSING THE 600 METER MARK. NOTHING MOVING HERE OTHER THAN US.
WE'LL KEEP ON HEADING TO THE CENTER.
HOW BEAUTIFUL...
VWAN
SO QUIET ...
PING
GRRM
YEAH.
GWOOM
IT'S LIKE A CORAL REEF THAT'S BEEN GROWN TO A BLUEPRINT.
VRRR
!
VWAN
LOOK DOWN.
AT THE CREVICE, THE COLOR IS DIFFERENT FROM THE SURROUNDINGS.
BIP BIP
BLIP
BLIP

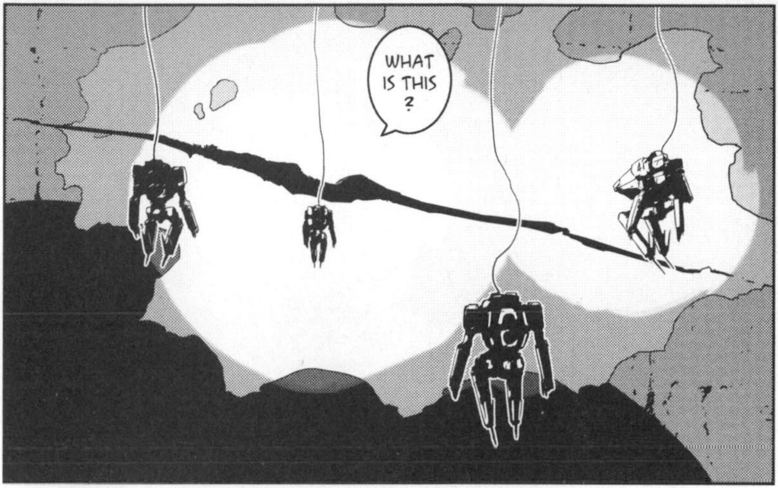
WHAT IS THIS ?

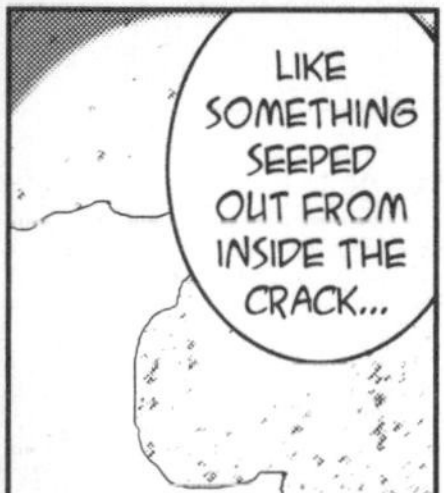
LIKE SOMETHING SEEPED OUT FROM INSIDE THE CRACK...

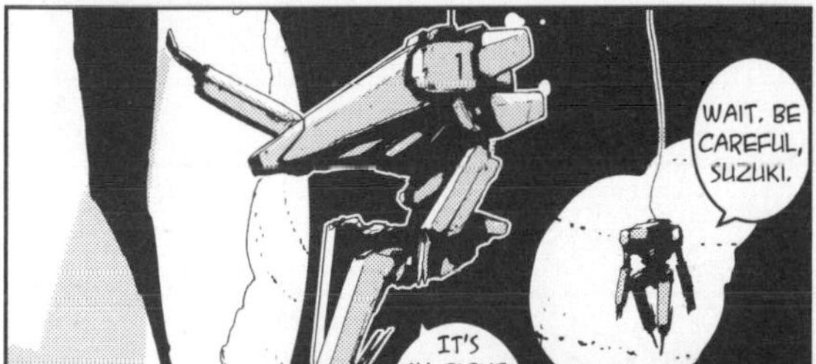
WAIT. BE CAREFUL, SUZUKI.
IT'S ALL RIGHT, KOBAYASHI. JUST A RESEARCH MISSION.

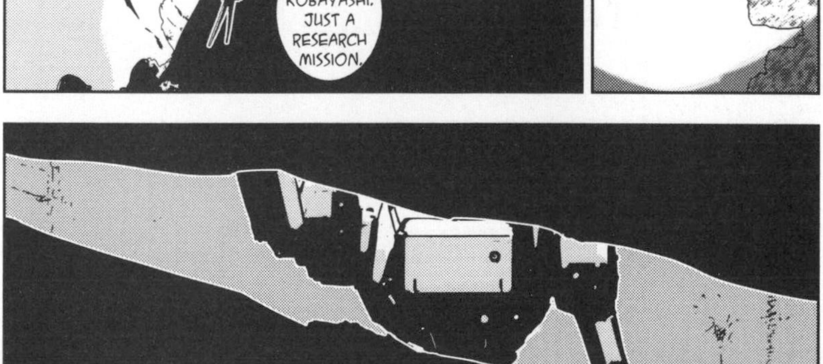

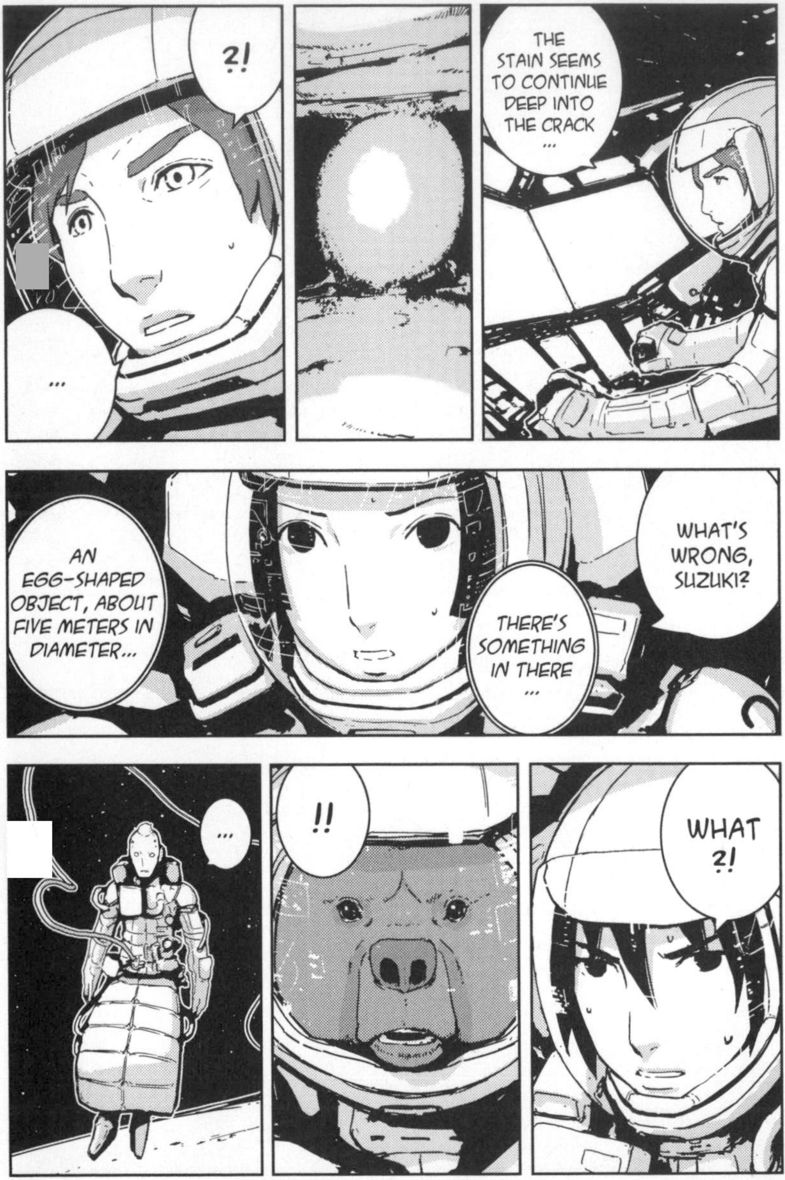
THE STAIN SEEMS TO CONTINUE DEEP INTO THE CRACK ...
?!
...
WHAT'S WRONG, SUZUKI?
THERE'S SOMETHING IN THERE ...
AN EGG-SHAPED OBJECT, ABOUT FIVE METERS IN DIAMETER...
WHAT ?!
!!
...

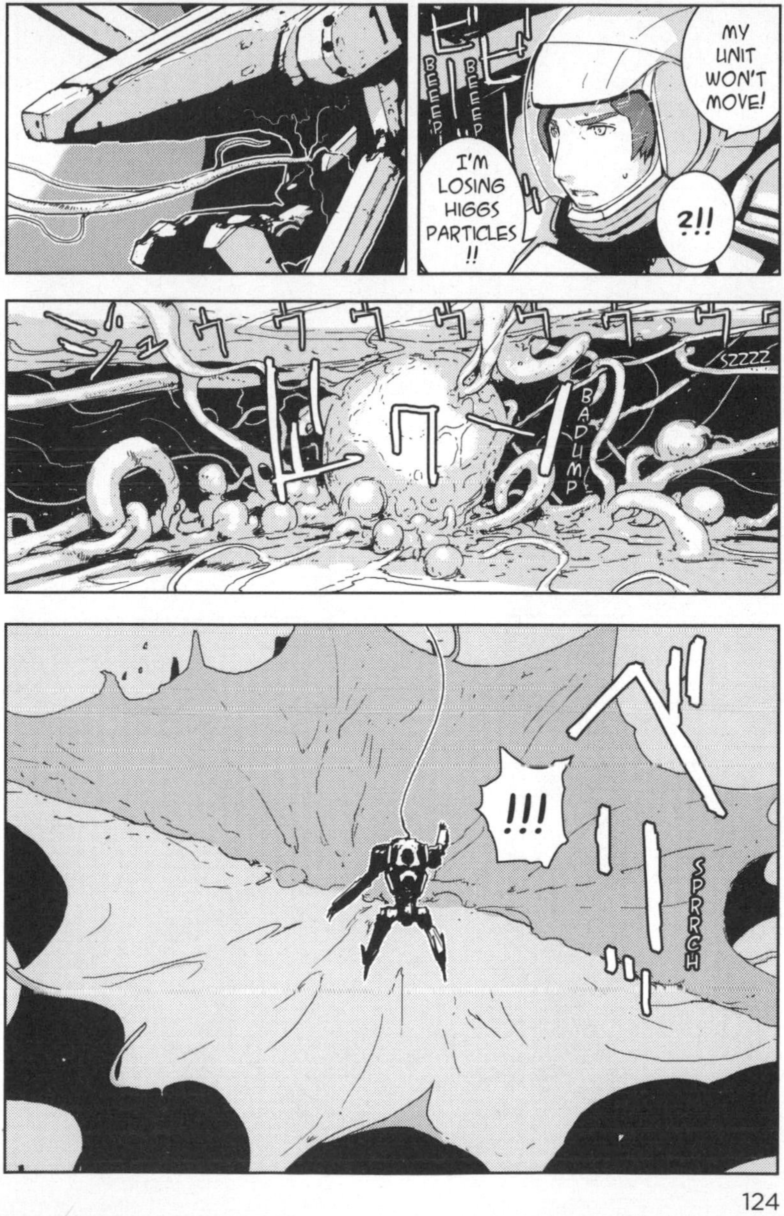
MY UNIT WON'T MOVE!
BEEEP
BEEEP
I'M LOSING HIGGS PARTICLES !!
?!!
SZZZZ
BADUMP
SPRRCH
!!!

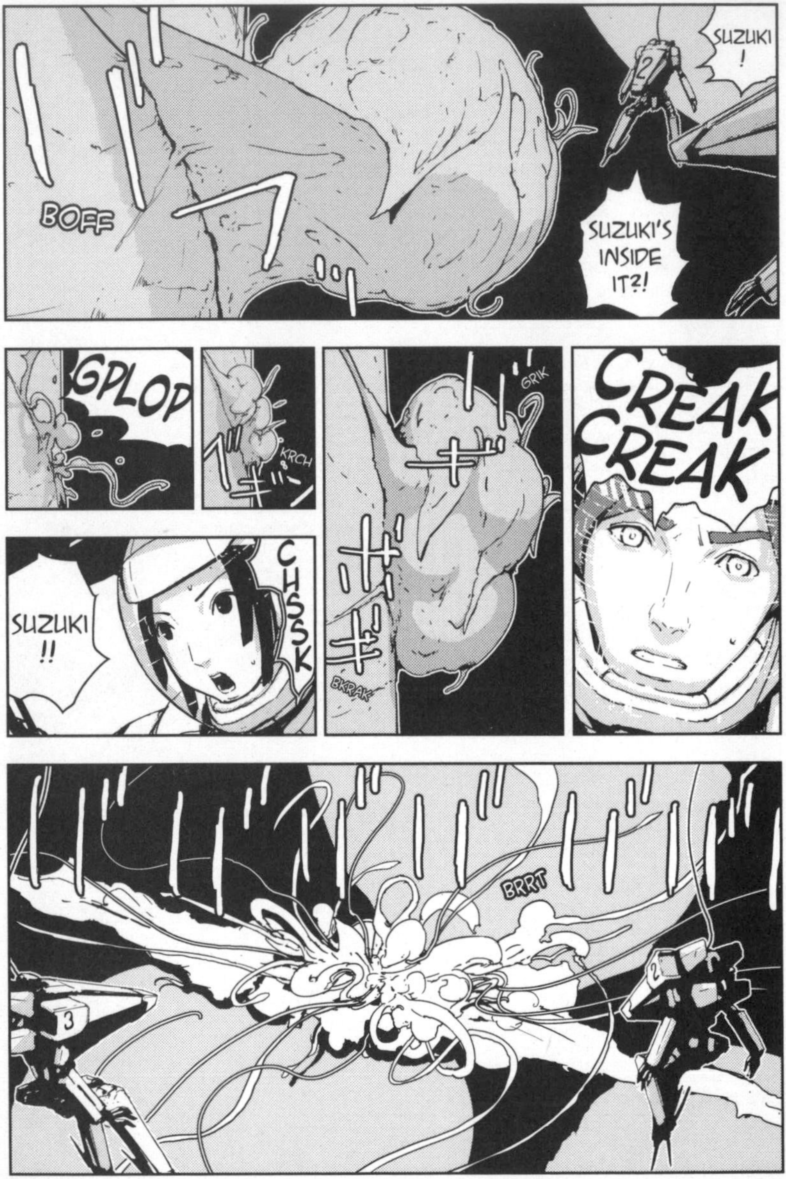
SUZUKI !
SUZUKI'S INSIDE IT?!
BOFF
CREAK CREAK
GRIK
BKRAK
KRCH
GPLOP
CHSSK
SUZUKI !!
BRRT

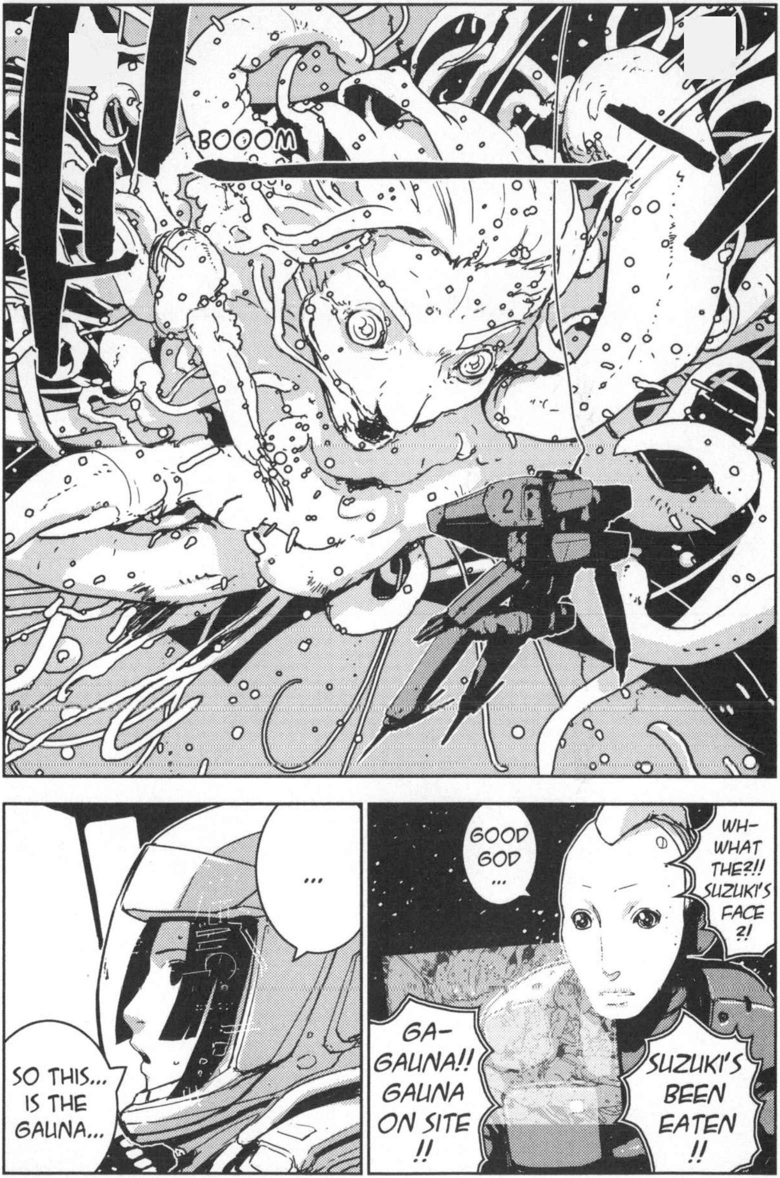
BOOOM
WH- WHAT THE?!! SUZUKI'S FACE ?!
GOOD GOD ...
SUZUKI'S BEEN EATEN !!
GA- GAUNA!! GAUNA ON SITE !!
...
SO THIS... IS THE GAUNA...

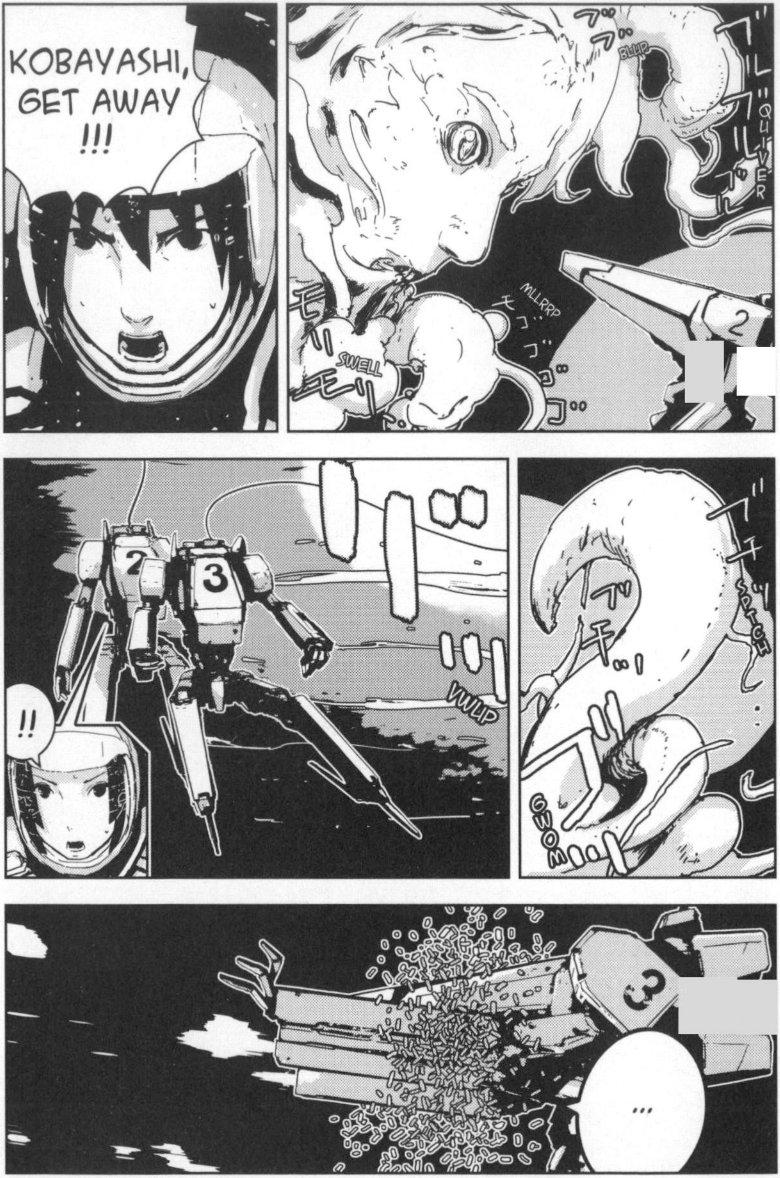
BLUP
QUIVER
MLLRRP
SWELL
KOBAYASHI, GET AWAY !!!
SPLTCH
GWOM
VWLIP
!!
...

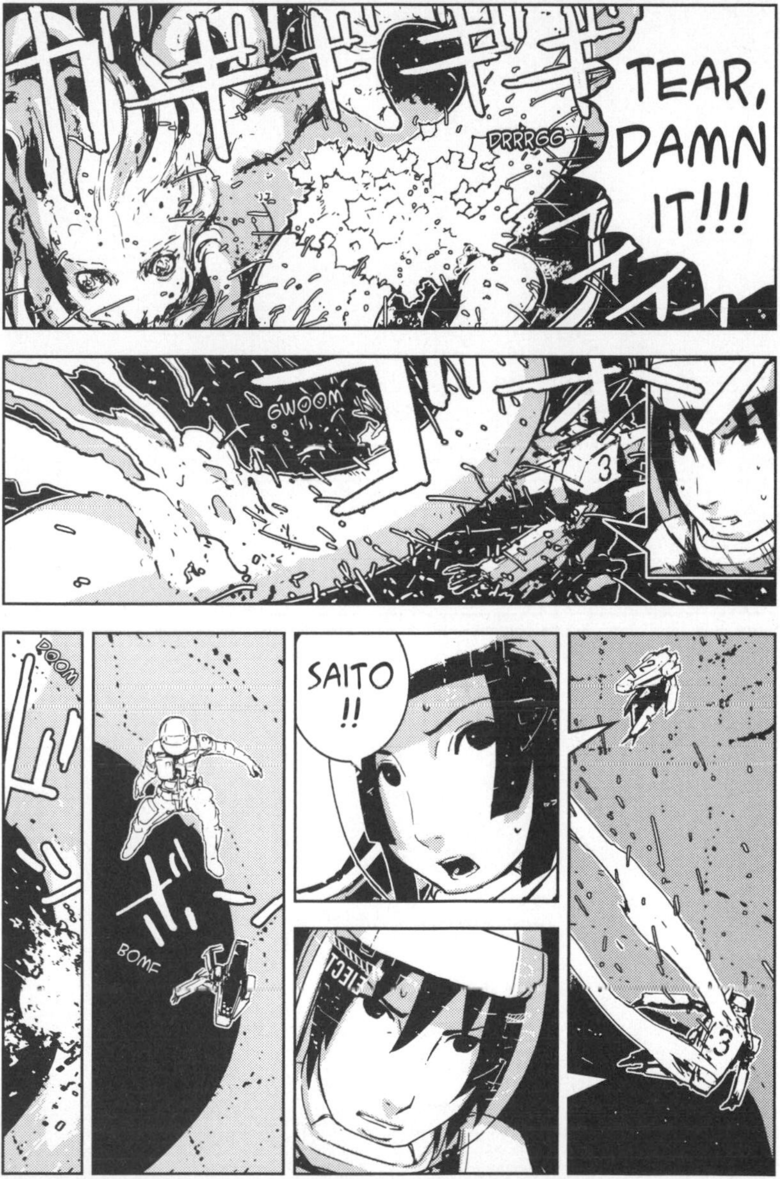
TEAR, DAMN IT!!!
DRRRGG
GWOOM
SAITO !!
BOMF
DOOM

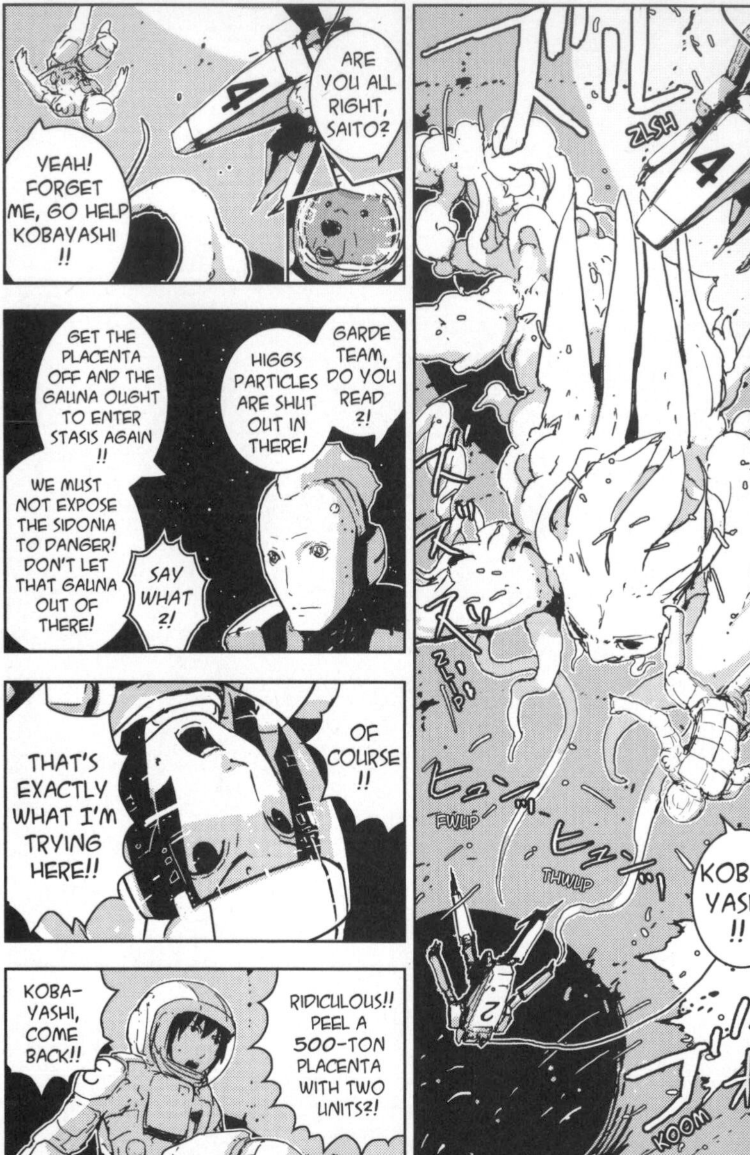
ZLSH
ZLIP
FWLIP
THWLP
KOBA-YASHI!!
KOOM
ARE YOU ALL RIGHT, SAITO?
YEAH! FORGET ME, GO HELP KOBAYASHI!!
GARDE TEAM, DO YOU READ?!
HIGGS PARTICLES ARE SHUT OUT IN THERE!
GET THE PLACENTA OFF AND THE GAUNA OUGHT TO ENTER STASIS AGAIN!!
WE MUST NOT EXPOSE THE SIDONIA TO DANGER! DON'T LET THAT GAUNA OUT OF THERE!
SAY WHAT?!
OF COURSE!!
THAT'S EXACTLY WHAT I'M TRYING HERE!!
RIDICULOUS!! PEEL A 500-TON PLACENTA WITH TWO UNITS?!
KOBA-YASHI, COME BACK!!

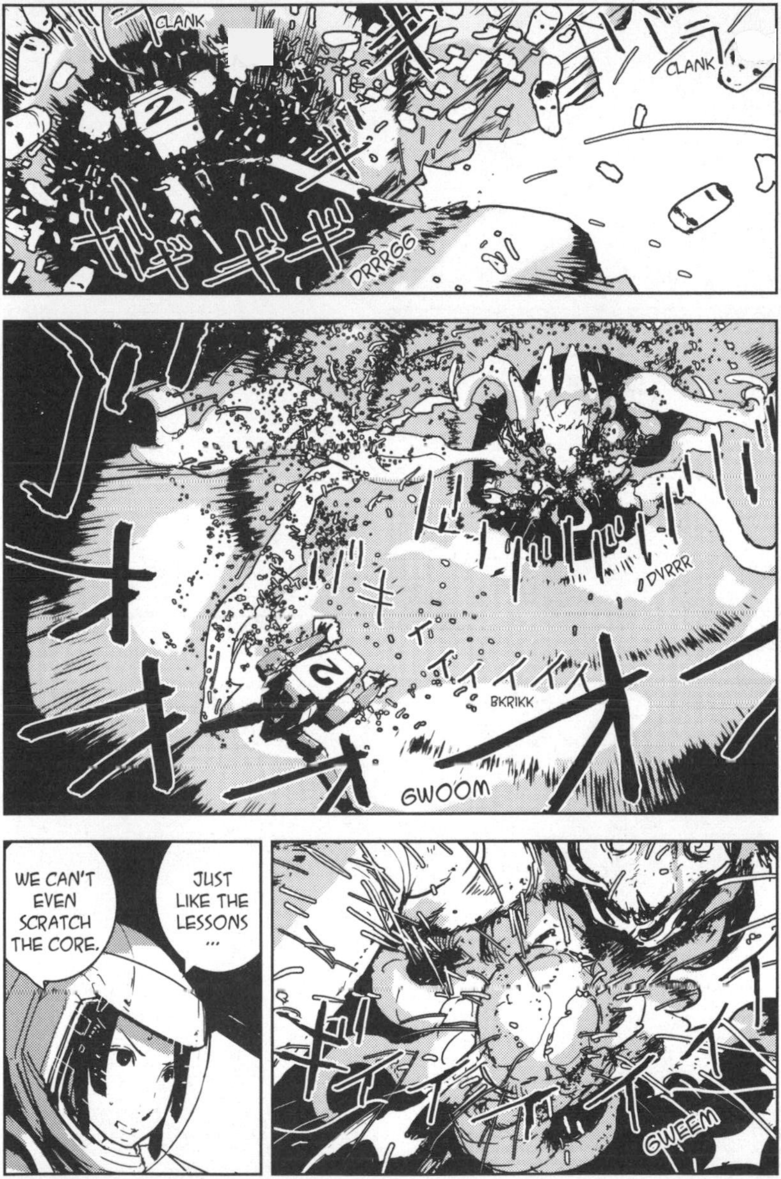
CLANK
CLANK
DRRRGG
DVRRR
BKRIKK
GWOOM
GWEEM
JUST LIKE THE LESSONS ...
WE CAN'T EVEN SCRATCH THE CORE.

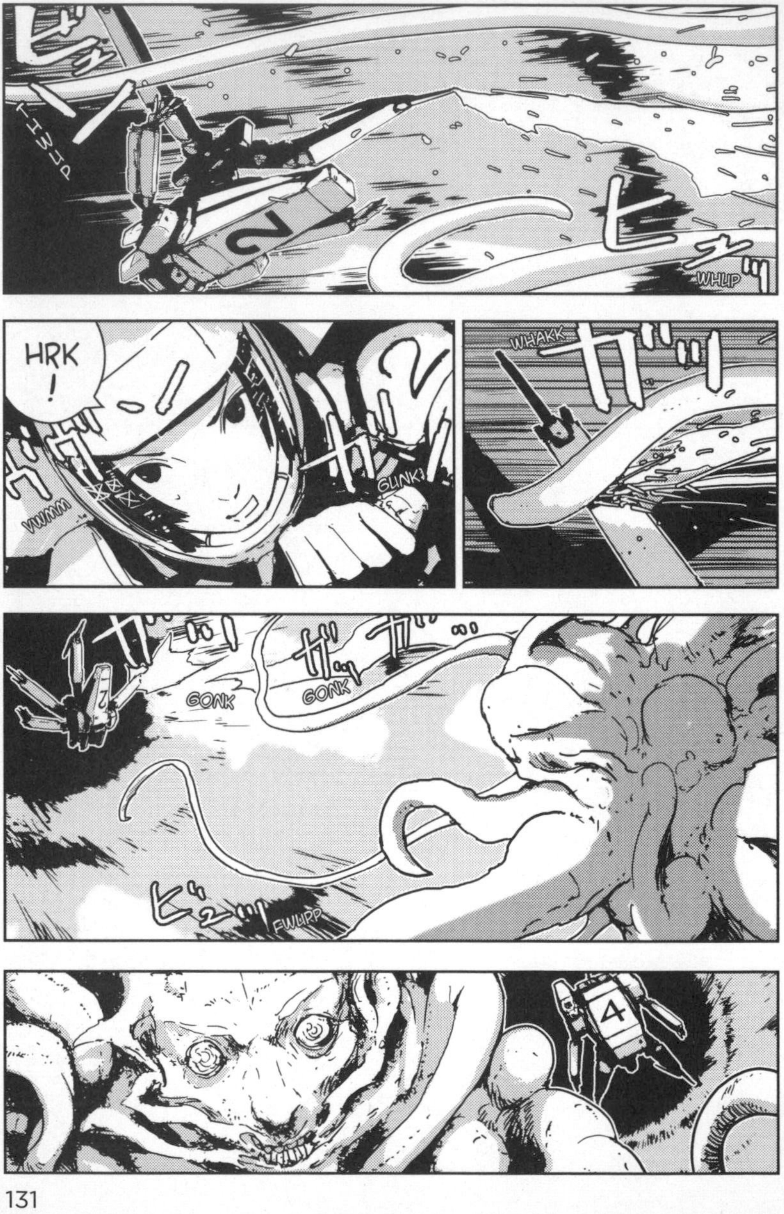

THWUP
WHLIP
WHAKK
HRK !
GUNK!
VWMM
GONK
GONK
FWUPP

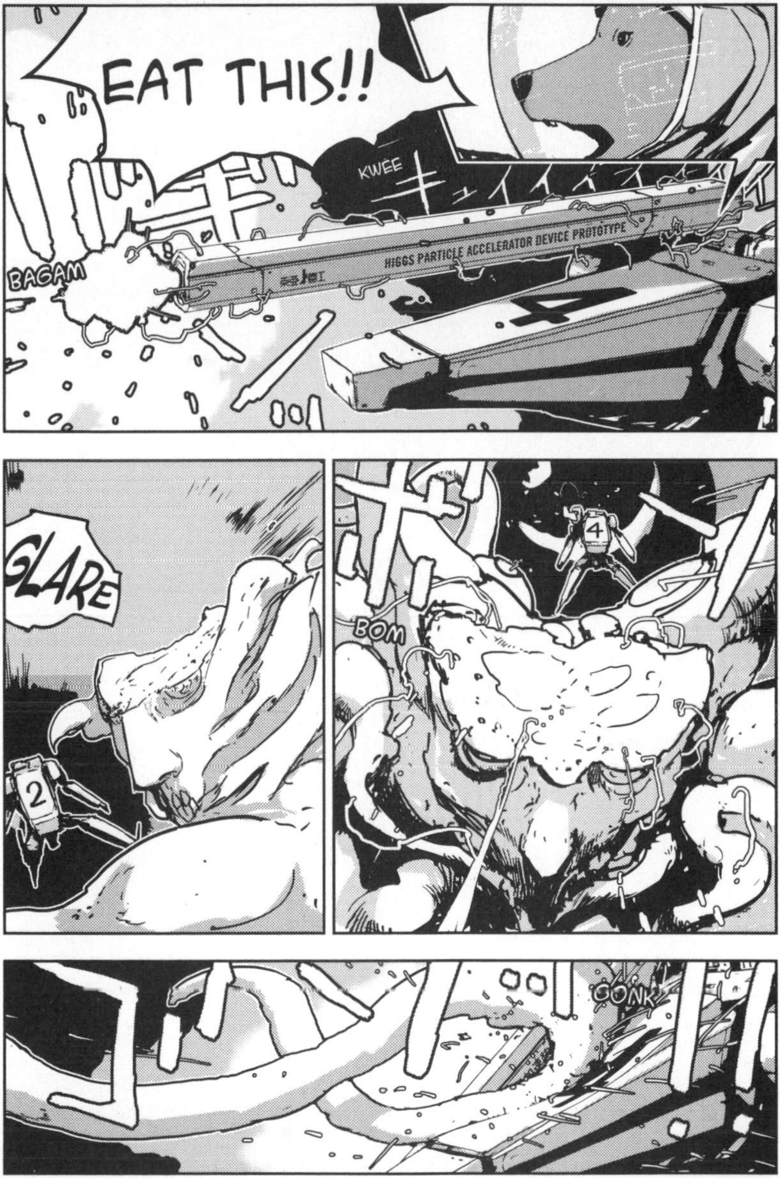
EAT THIS!!
KWEE
HIGGS PARTICLE ACCELERATOR DEVICE PROTOTYPE
BAGAM
BOM
GLARE
CONK

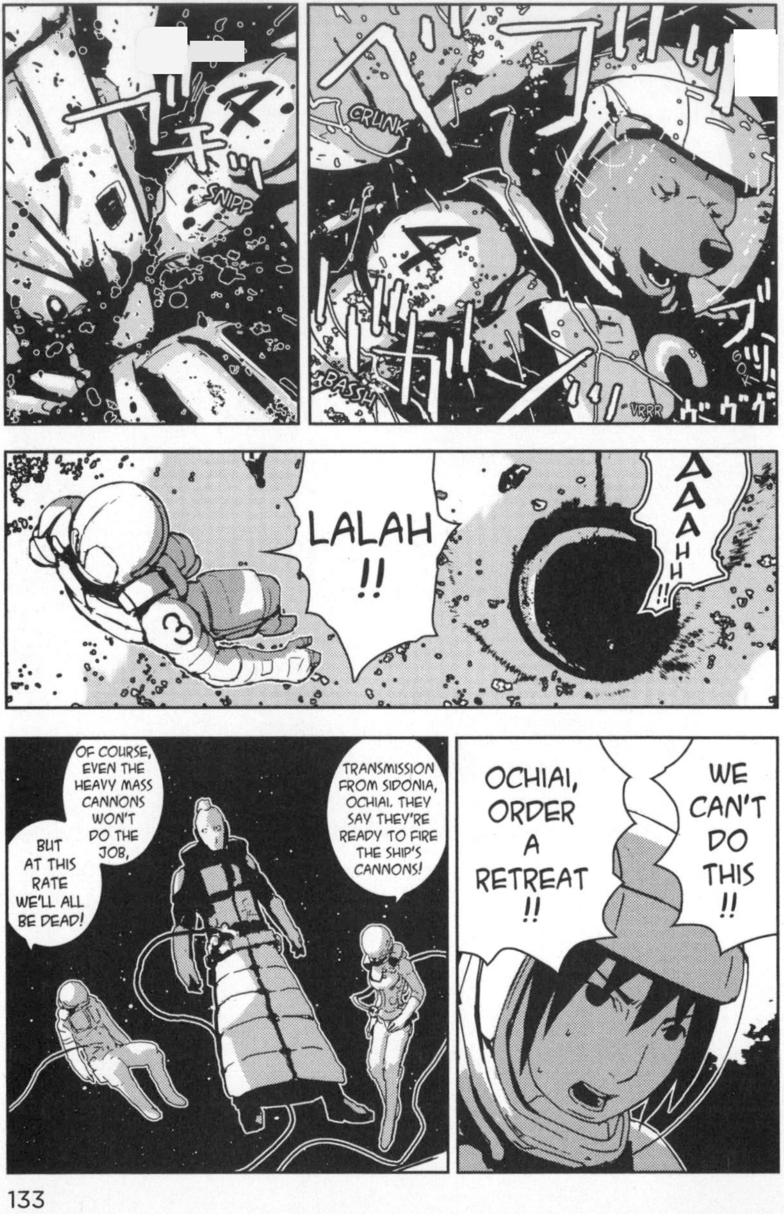
CRUNK
GOK
VRRR
BASH
SNIPP
AAAAHH!!
LALAH!!
WE CAN'T DO THIS!!
OCHIAI, ORDER A RETREAT!!
TRANSMISSION FROM SIDONIA, OCHIAI. THEY SAY THEY'RE READY TO FIRE THE SHIP'S CANNONS!
OF COURSE, EVEN THE HEAVY MASS CANNONS WON'T DO THE JOB,
BUT AT THIS RATE WE'LL ALL BE DEAD!

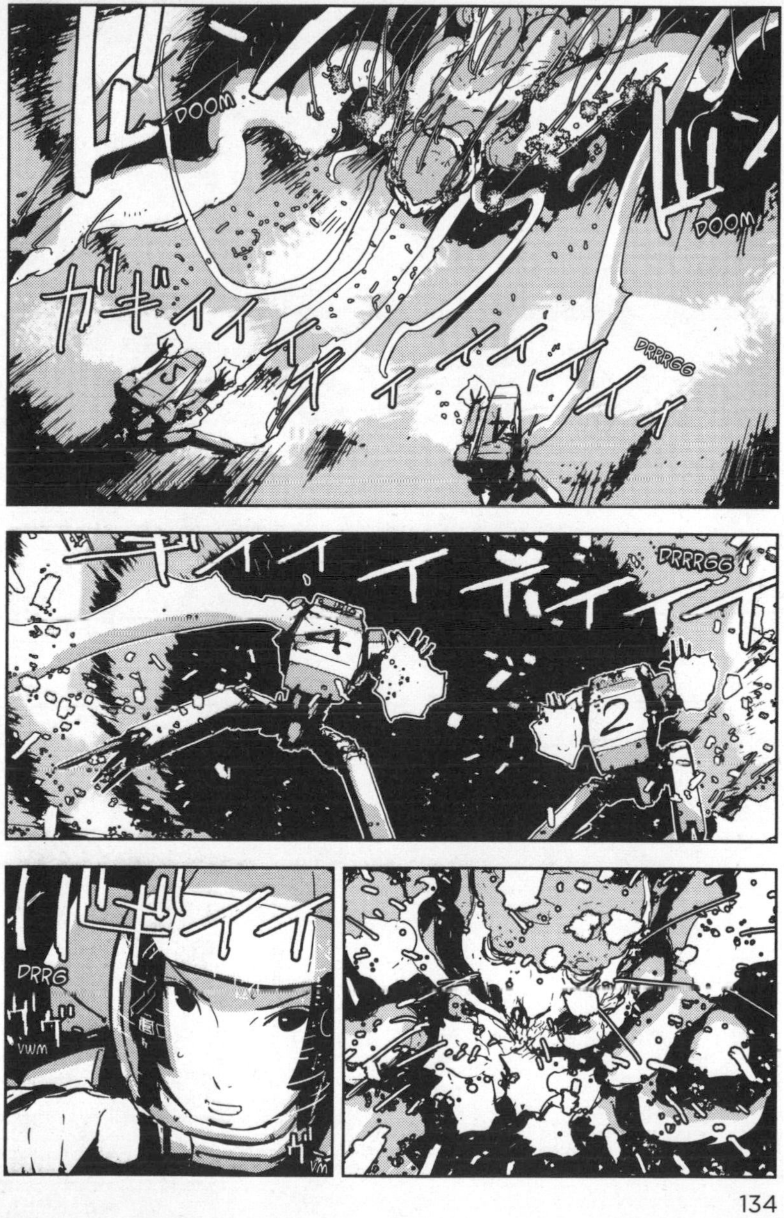
DOOM
DOOM
DRRRGG
DRRRGG
DRRG
VWM
VM

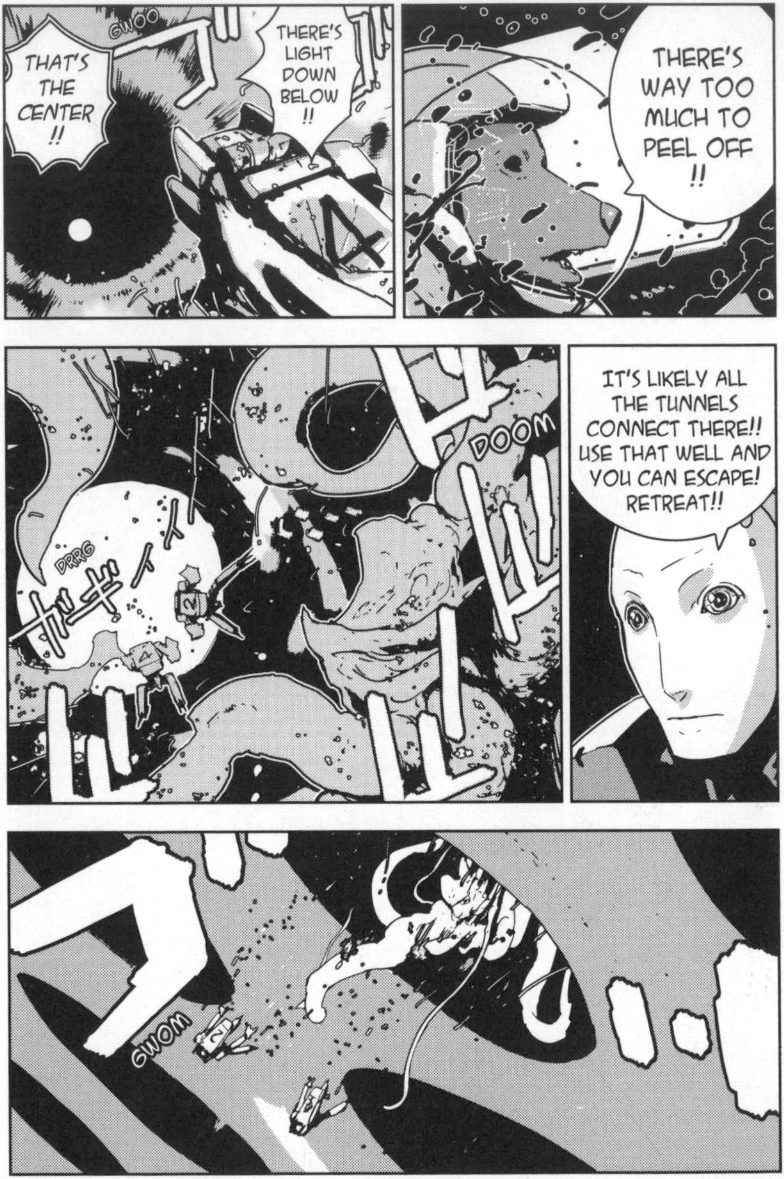
THERE'S WAY TOO MUCH TO PEEL OFF !!
THERE'S LIGHT DOWN BELOW !!
GWOO
THAT'S THE CENTER !!
IT'S LIKELY ALL THE TUNNELS CONNECT THERE!! USE THAT WELL AND YOU CAN ESCAPE! RETREAT!!
DOOM
DRRG
GWOM

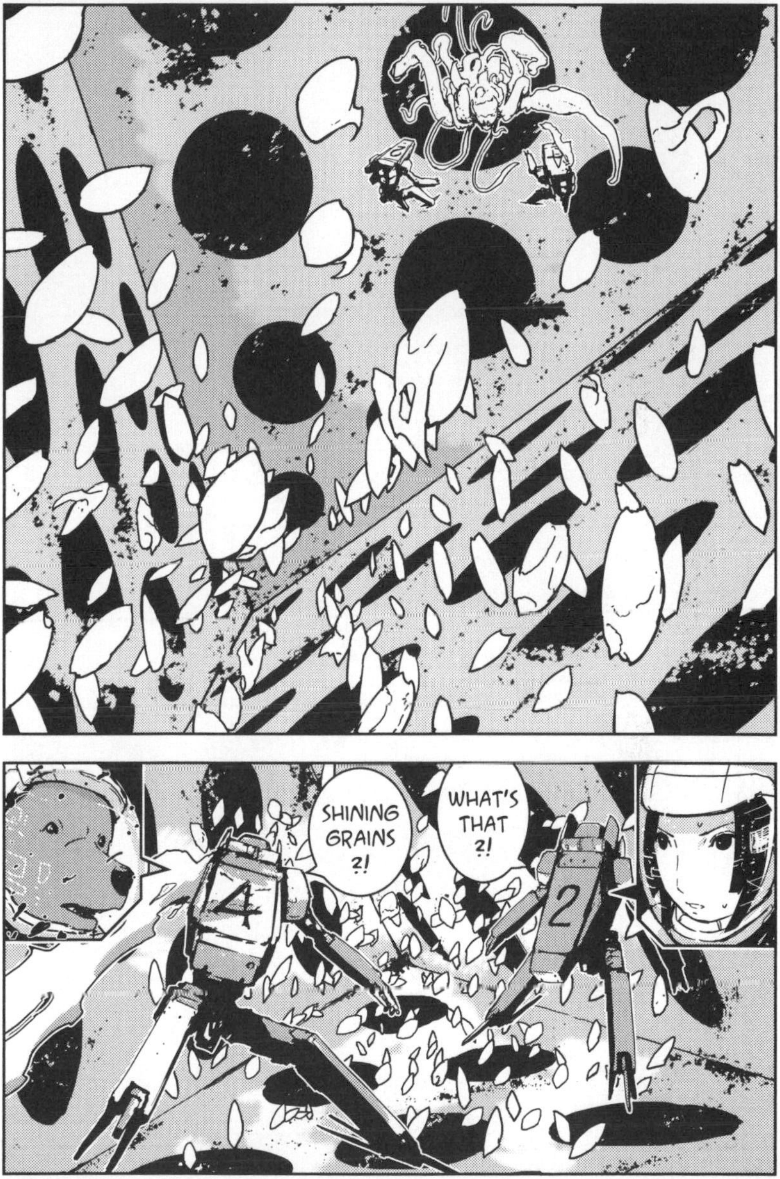
WHAT'S THAT ?!
SHINING GRAINS ?!

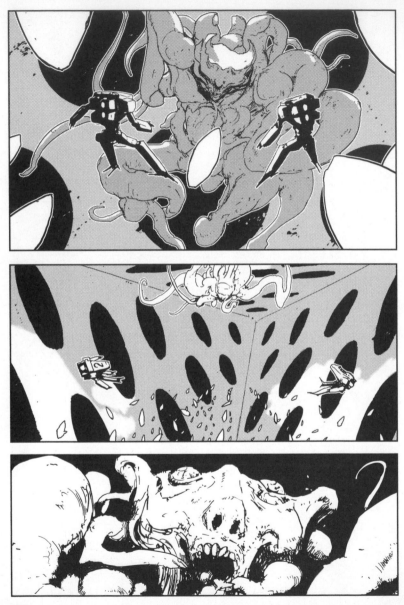

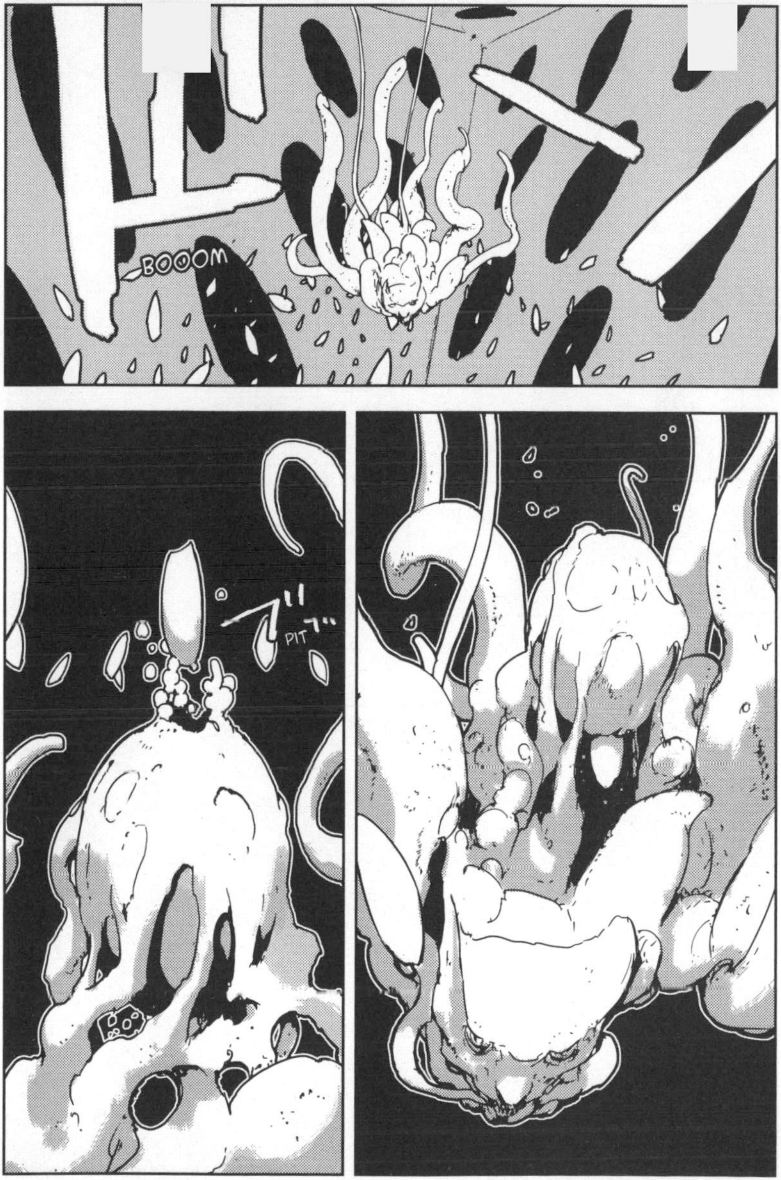
BOOOM
PIT

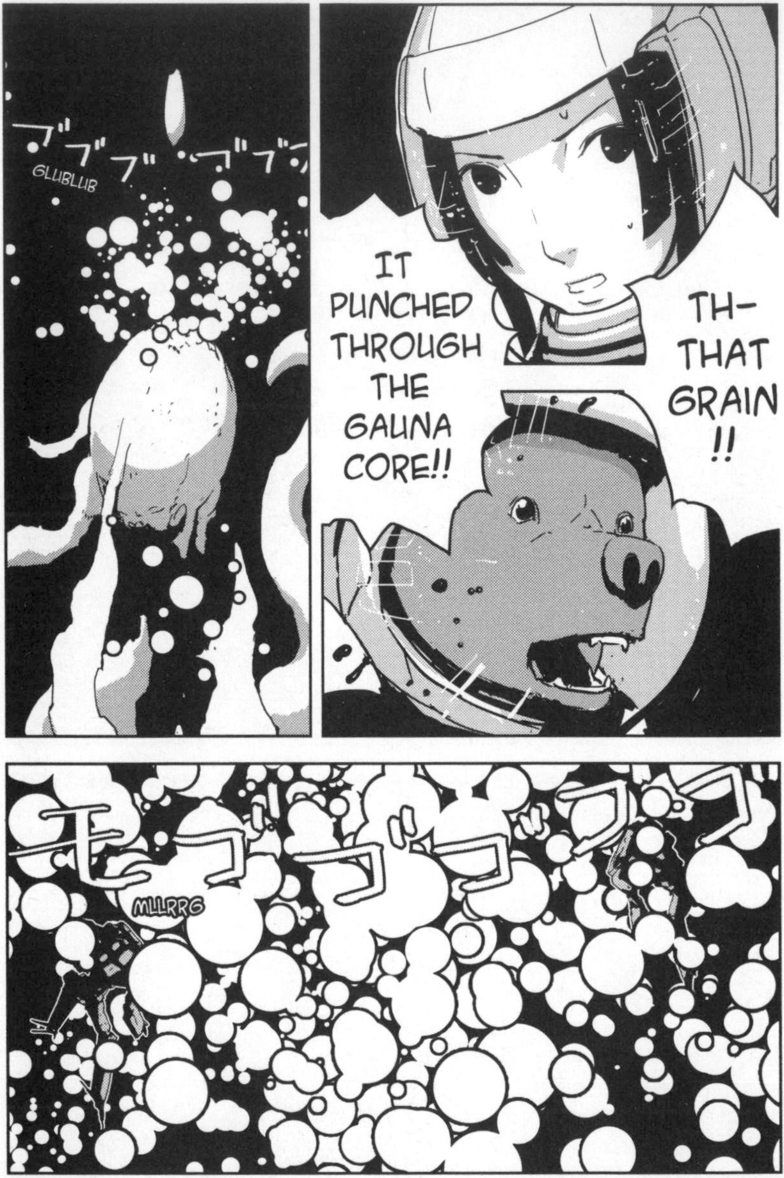
TH-THAT GRAIN!!
IT PUNCHED THROUGH THE GAUNA CORE!!
GLUBLUB
MLLLRRG

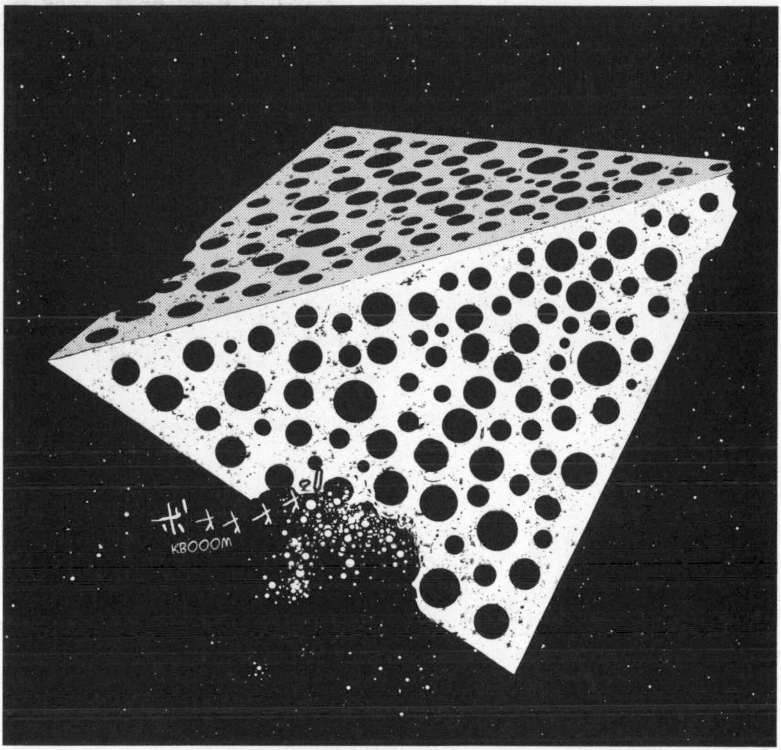
KBOOOM

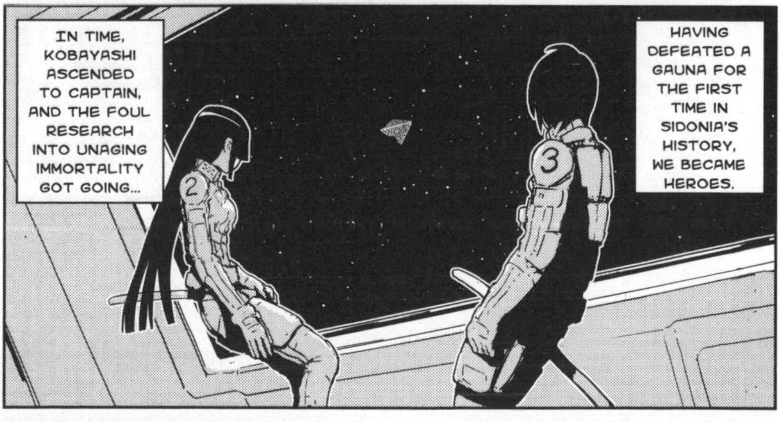
HAVING DEFEATED A GAUNA FOR THE FIRST TIME IN SIDONIA'S HISTORY, WE BECAME HEROES.
IN TIME, KOBAYASHI ASCENDED TO CAPTAIN, AND THE FOUL RESEARCH INTO UNAGING IMMORTALITY GOT GOING...
3
2

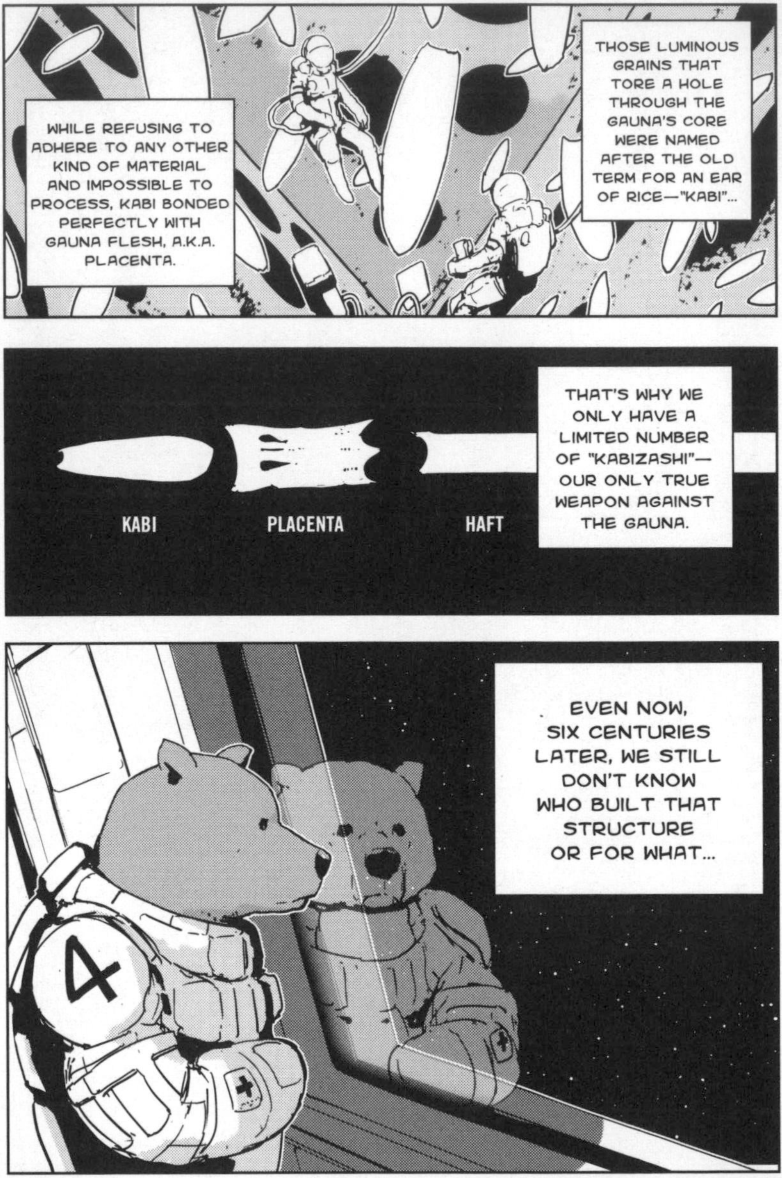
THOSE LUMINOUS GRAINS THAT TORE A HOLE THROUGH THE GAUNA'S CORE WERE NAMED AFTER THE OLD TERM FOR AN EAR OF RICE—"KABI"...
WHILE REFUSING TO ADHERE TO ANY OTHER KIND OF MATERIAL AND IMPOSSIBLE TO PROCESS, KABI BONDED PERFECTLY WITH GAUNA FLESH, A.K.A. PLACENTA.
THAT'S WHY WE ONLY HAVE A LIMITED NUMBER OF "KABIZASHI"— OUR ONLY TRUE WEAPON AGAINST THE GAUNA.
KABI
PLACENTA
HAFT
EVEN NOW, SIX CENTURIES LATER, WE STILL DON'T KNOW WHO BUILT THAT STRUCTURE OR FOR WHAT...
4

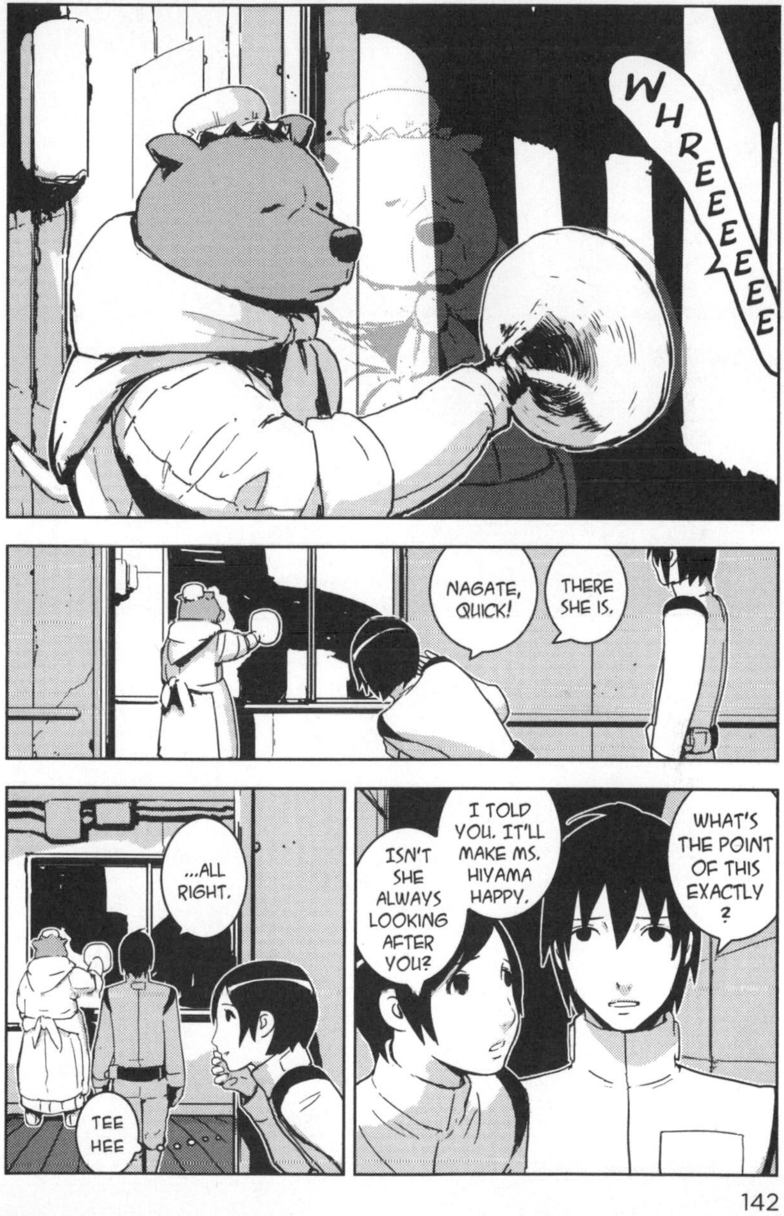
WHREEEEEE
THERE SHE IS.
NAGATE, QUICK!
WHAT'S THE POINT OF THIS EXACTLY?
I TOLD YOU. IT'LL MAKE MS. HIYAMA HAPPY.
ISN'T SHE ALWAYS LOOKING AFTER YOU?
...ALL RIGHT.
TEE HEE

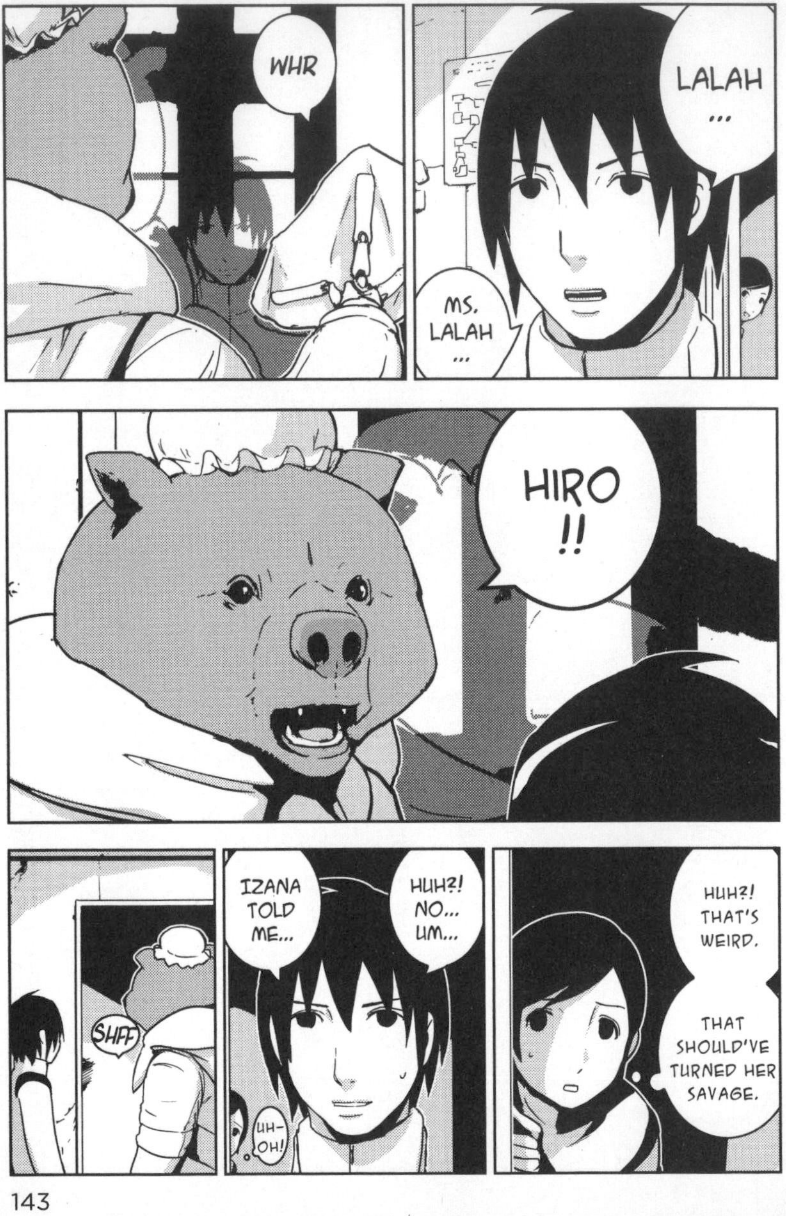
LALAH ...
MS. LALAH ...
WHR
HIRO !!
HUH?! THAT'S WEIRD.
THAT SHOULD'VE TURNED HER SAVAGE.
HUH?! NO... UM...
IZANA TOLD ME...
UH-OH!
SHFF

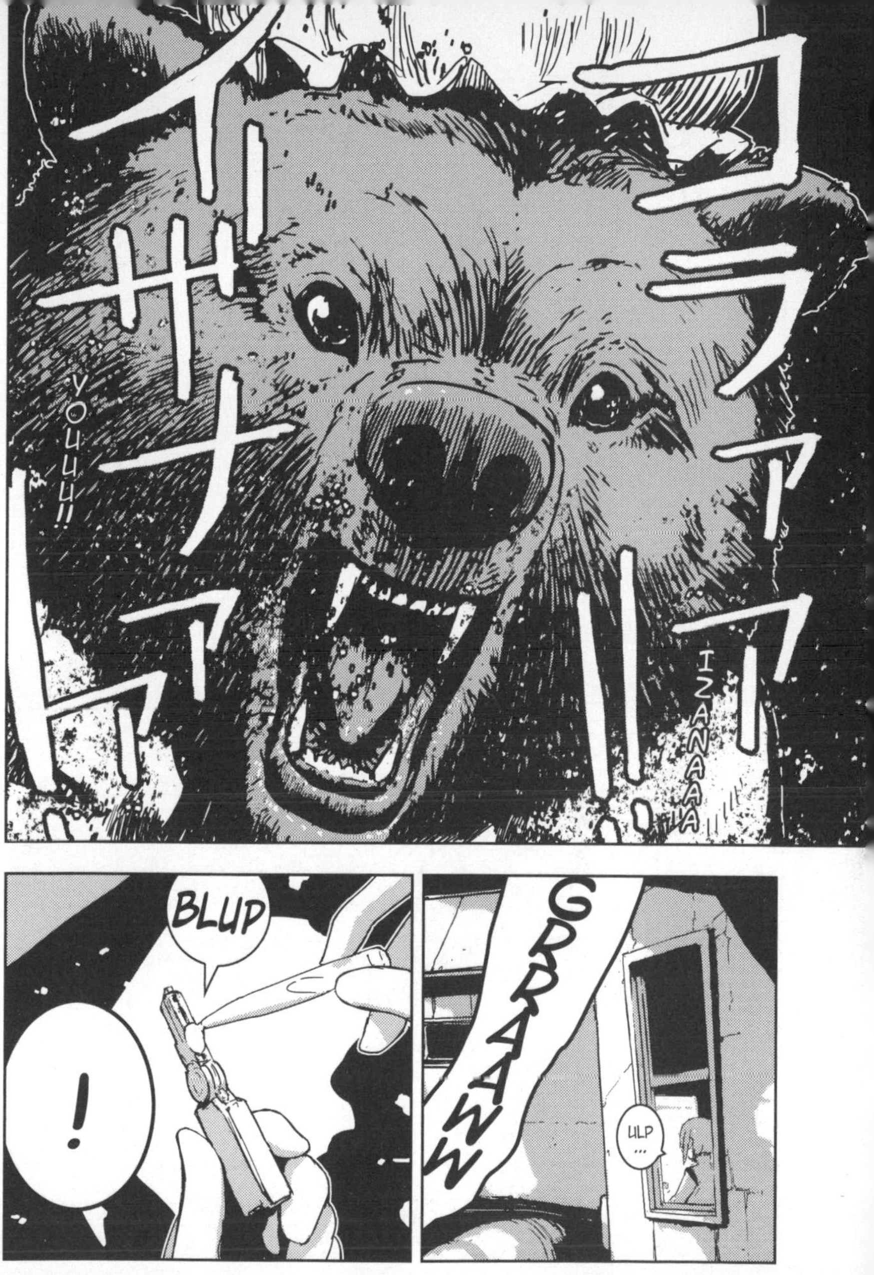

コラァァー!!
IZANAAAA
イーザナァァー!!
YOUUUU!!
GRRAAWW
ULP...
BLUP
!

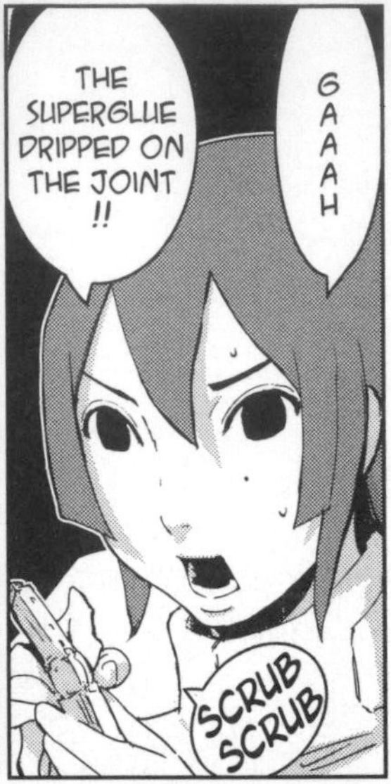
GAAAH
THE SUPERGLUE DRIPPED ON THE JOINT !!
SCRUB SCRUB

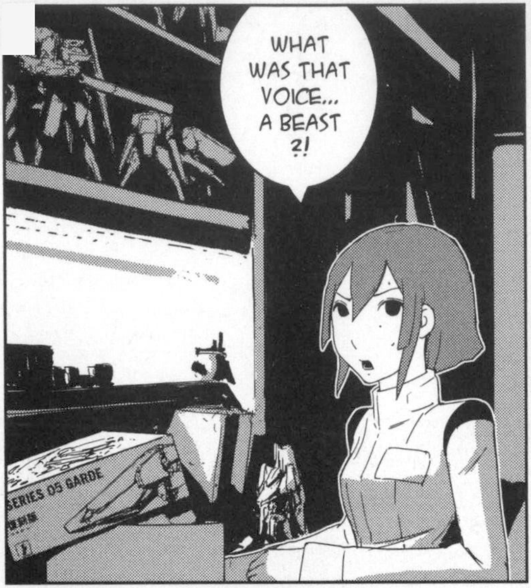
WHAT WAS THAT VOICE... A BEAST ?!
SERIES 05 GARDE

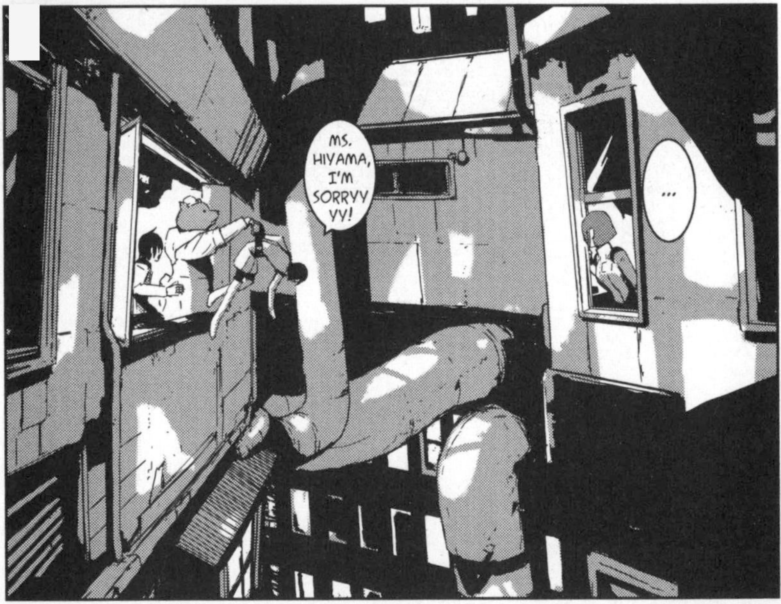
...
MS. HIYAMA, I'M SORRYY YY!

Chapter 14: END

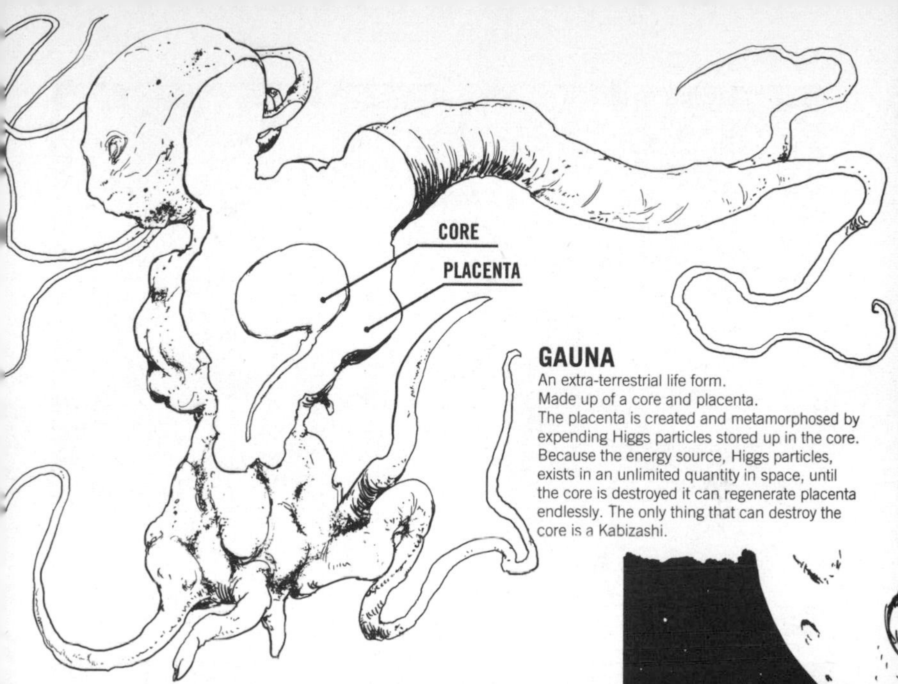
CORE
PLACENTA

GAUNA

An extra-terrestrial life form.
Made up of a core and placenta.
The placenta is created and metamorphosed by expending Higgs particles stored up in the core. Because the energy source, Higgs particles, exists in an unlimited quantity in space, until the core is destroyed it can regenerate placenta endlessly. The only thing that can destroy the core is a Kabizashi.

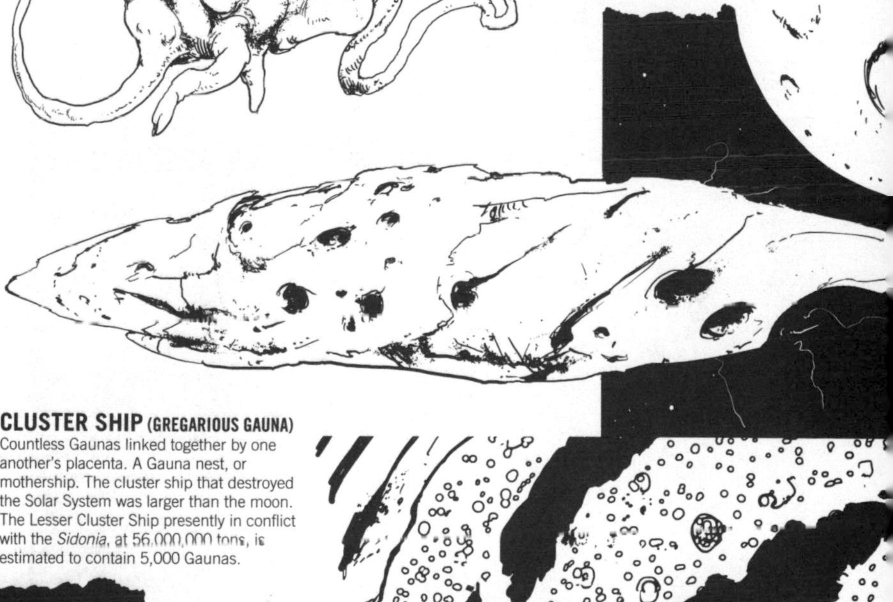

CLUSTER SHIP (GREGARIOUS GAUNA)

Countless Gaunas linked together by one another's placenta. A Gauna nest, or mothership. The cluster ship that destroyed the Solar System was larger than the moon. The Lesser Cluster Ship presently in conflict with the *Sidonia*, at 56,000,000 tons, is estimated to contain 5,000 Gaunas.

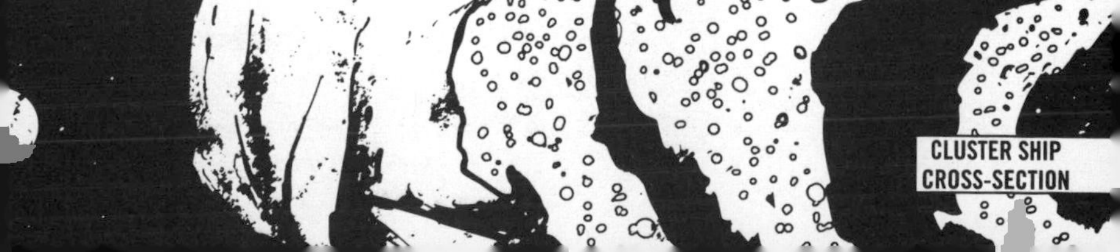
CLUSTER SHIP
CROSS-SECTION

Chapter 15: Hiroki Saito's Spree

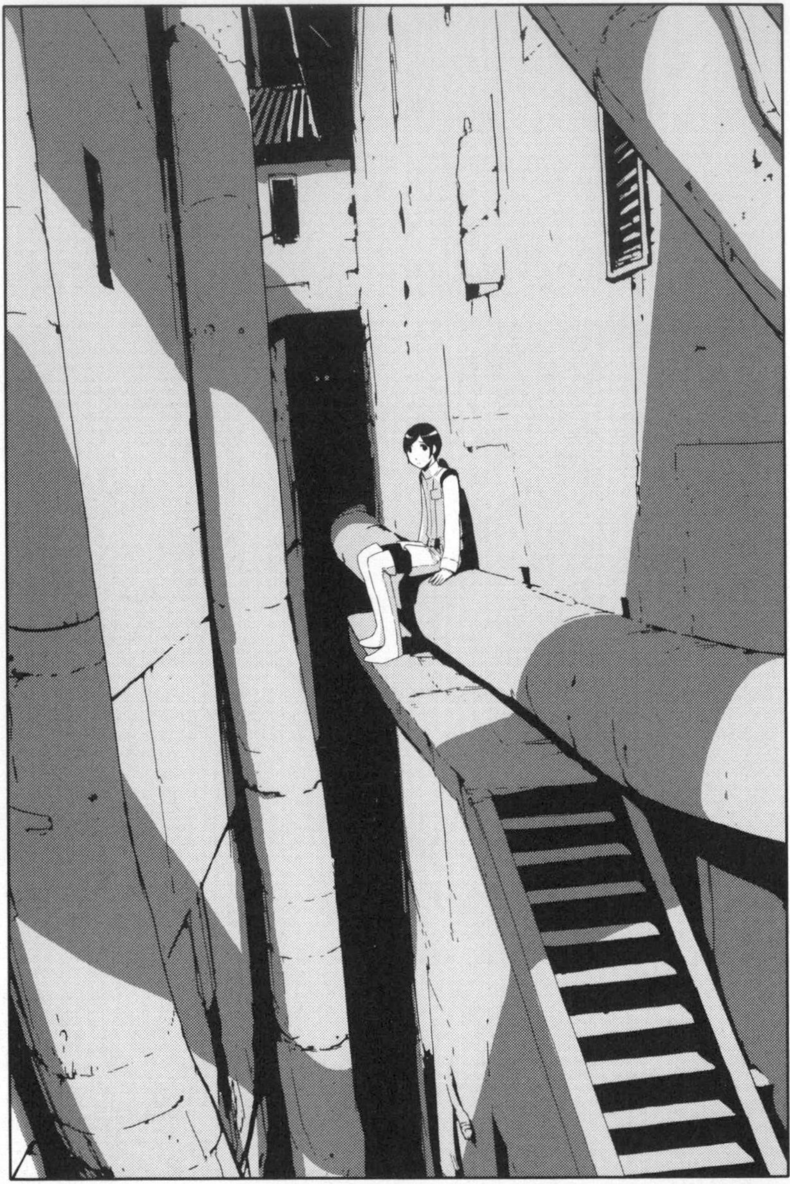

One Hundred Sights of Sidonia Part Twelve:
Pilot Dormitories Rear

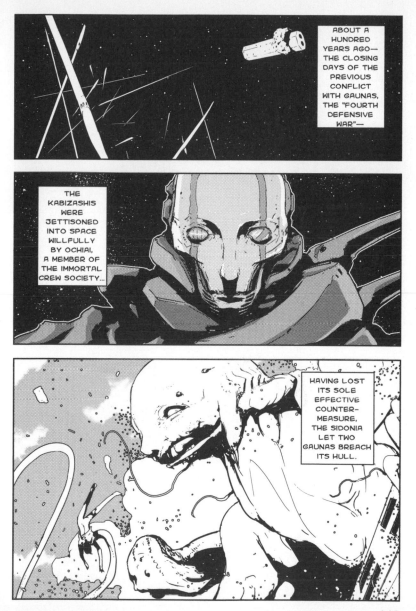

ABOUT A HUNDRED YEARS AGO— THE CLOSING DAYS OF THE PREVIOUS CONFLICT WITH GAUNAS, THE "FOURTH DEFENSIVE WAR"—

THE KABIZASHIS WERE JETTISONED INTO SPACE WILLFULLY BY OCHIAI, A MEMBER OF THE IMMORTAL CREW SOCIETY...

HAVING LOST ITS SOLE EFFECTIVE COUNTER-MEASURE, THE SIDONIA LET TWO GAUNAS BREACH ITS HULL.

148

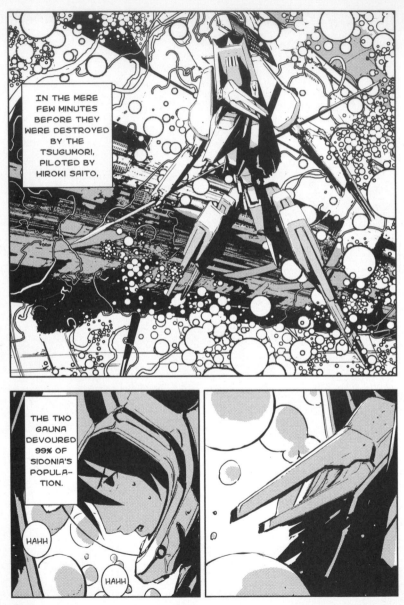

IN THE MERE FEW MINUTES BEFORE THEY WERE DESTROYED BY THE TSUGUMORI, PILOTED BY HIROKI SAITO,

THE TWO GAUNA DEVOURED 99% OF SIDONIA'S POPULATION.

HAHH

HAHH

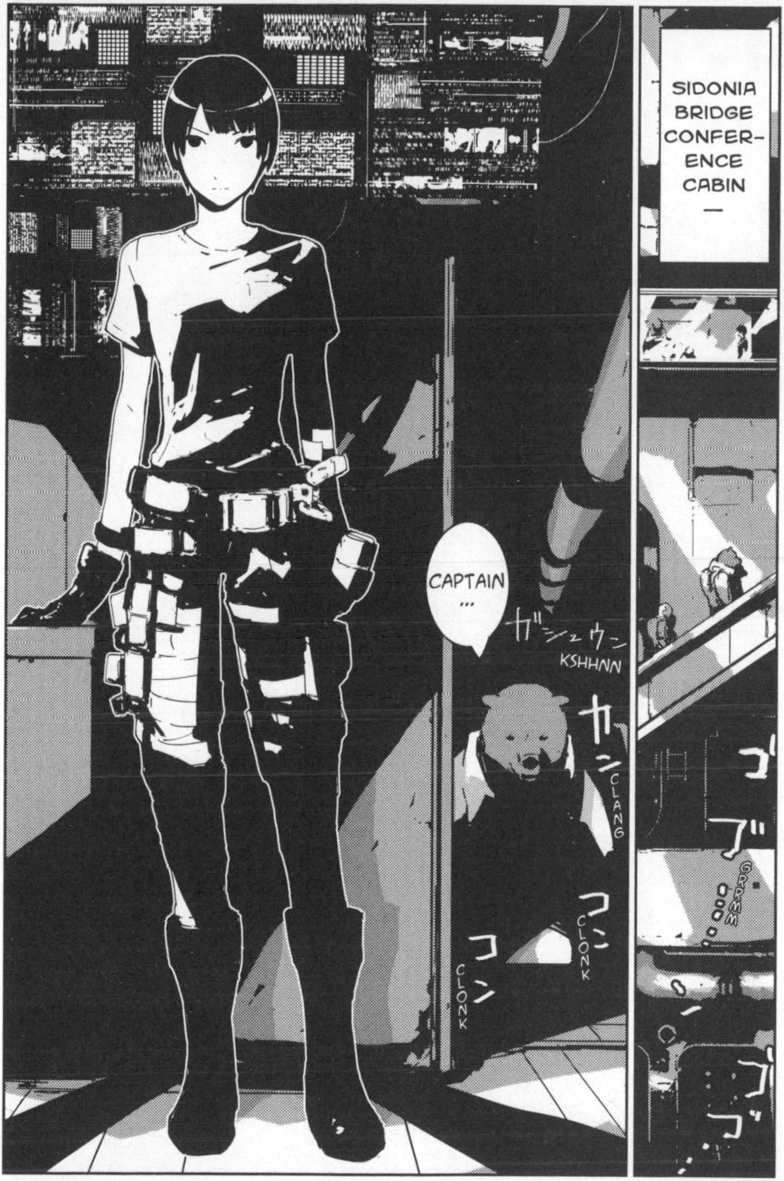
SIDONIA BRIDGE CONFER- ENCE CABIN —
GRPMM
CAPTAIN ...
ガシュウーン
KSHHNN
カン
CLANG
コン
CLONK
コン
CLONK

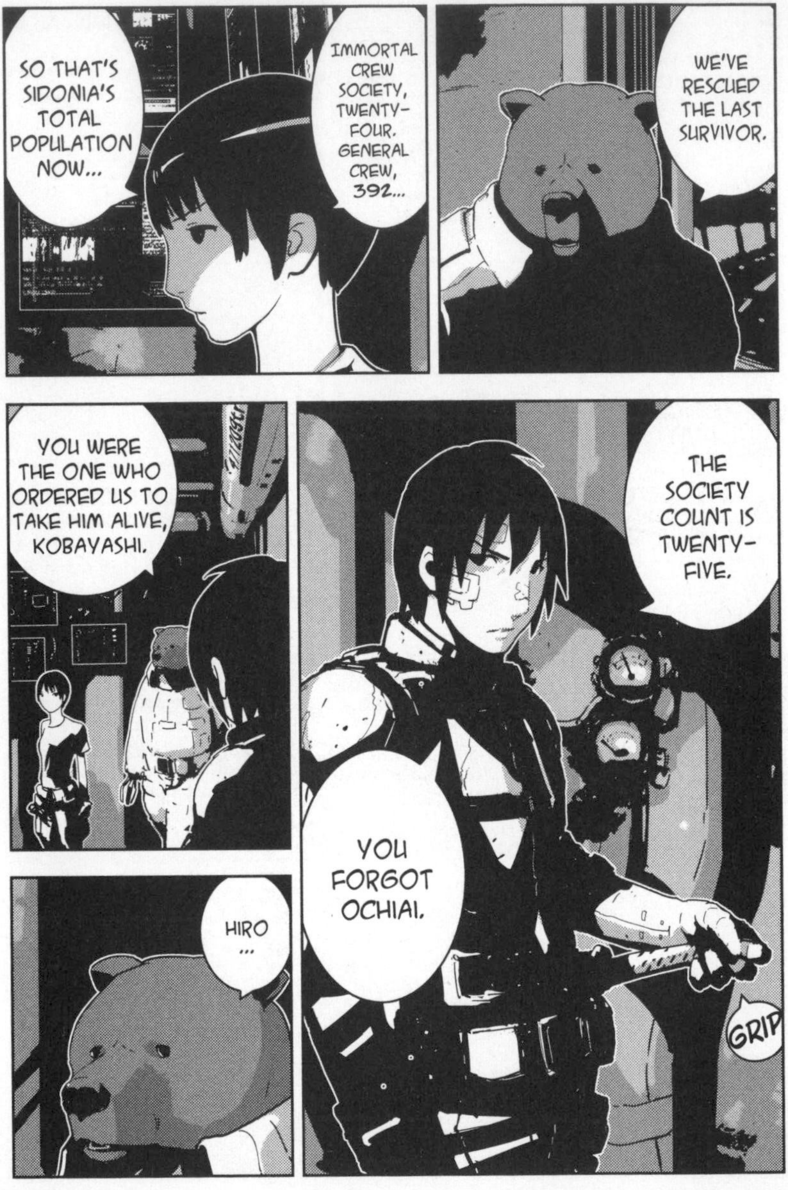
WE'VE RESCUED THE LAST SURVIVOR.
IMMORTAL CREW SOCIETY, TWENTY-FOUR. GENERAL CREW, 392...
SO THAT'S SIDONIA'S TOTAL POPULATION NOW...
THE SOCIETY COUNT IS TWENTY-FIVE.
YOU FORGOT OCHIAI.
GRIP
YOU WERE THE ONE WHO ORDERED US TO TAKE HIM ALIVE, KOBAYASHI.
HIRO ...

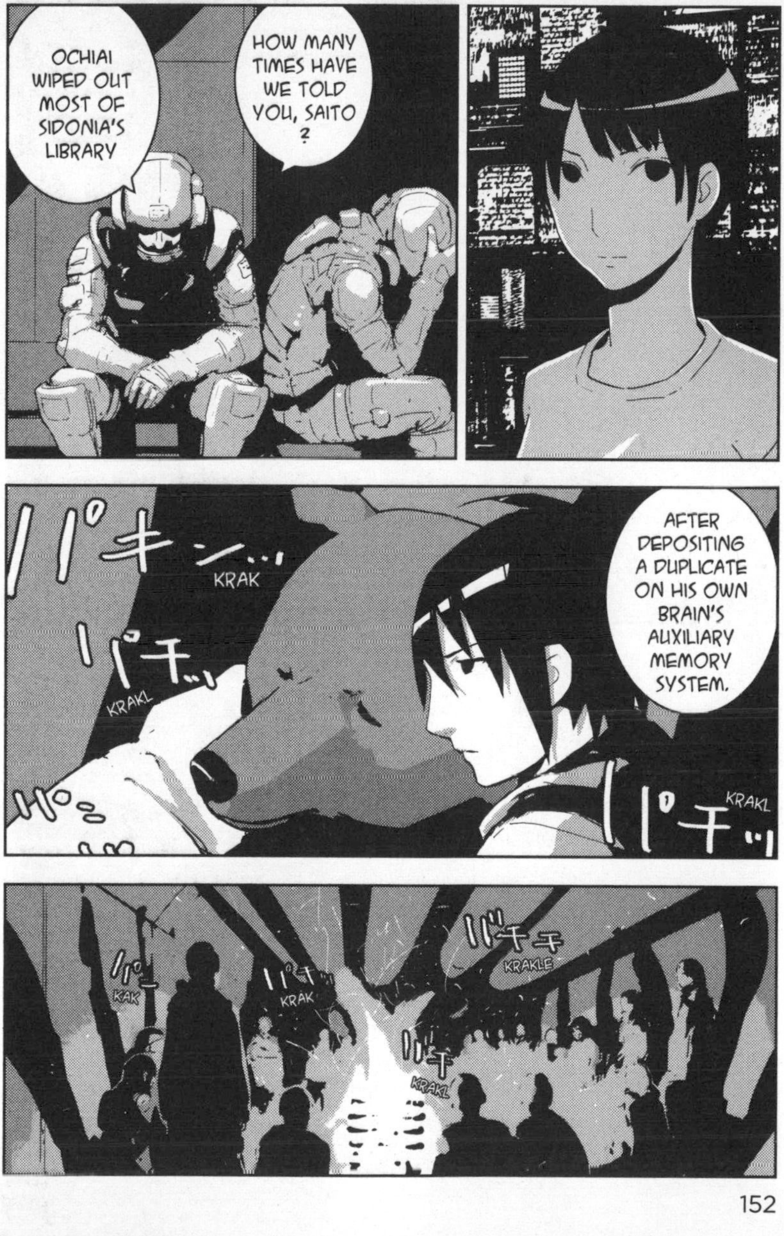
HOW MANY TIMES HAVE WE TOLD YOU, SAITO ?
OCHIAI WIPED OUT MOST OF SIDONIA'S LIBRARY
AFTER DEPOSITING A DUPLICATE ON HIS OWN BRAIN'S AUXILIARY MEMORY SYSTEM.
KRAK
KRAKL
KRAKL
KRAKLE
KRAK
KAK
KRAKL

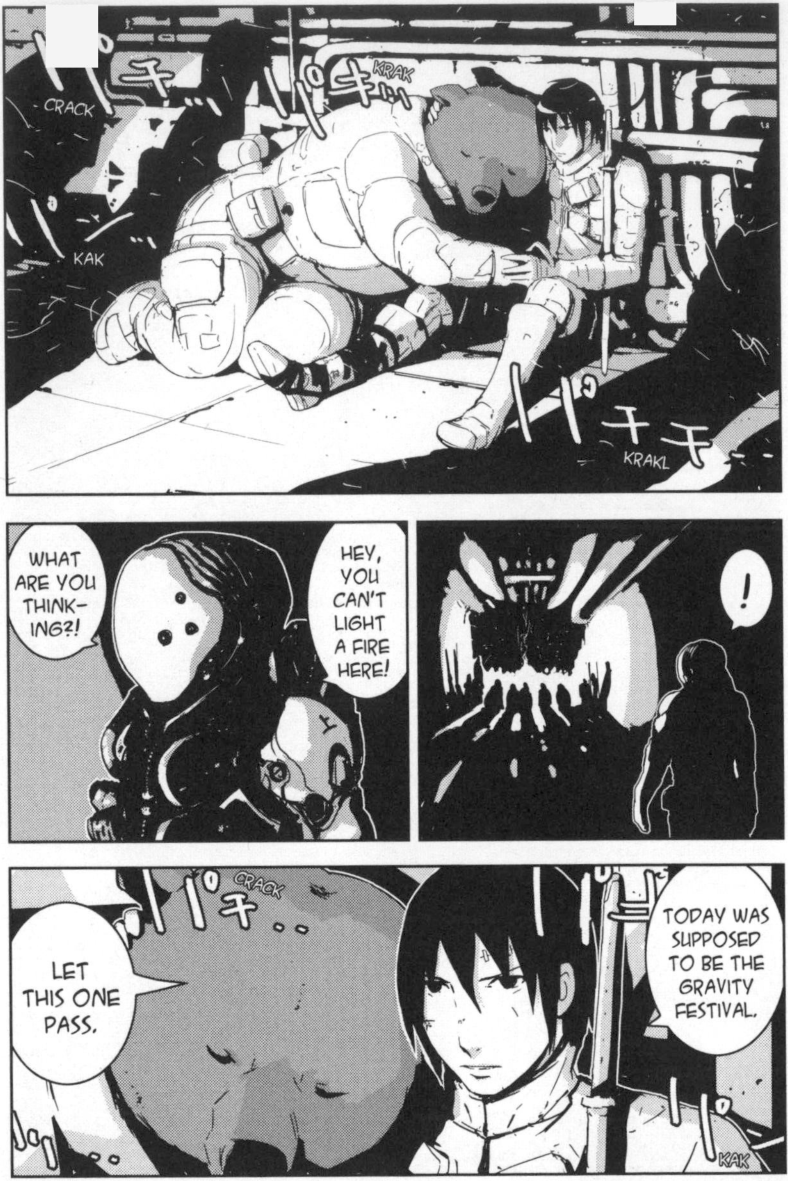
KRAK
CRACK
KAK
KRAKL
!
HEY, YOU CAN'T LIGHT A FIRE HERE!
WHAT ARE YOU THINKING?!
CRACK
TODAY WAS SUPPOSED TO BE THE GRAVITY FESTIVAL.
LET THIS ONE PASS.
KAK

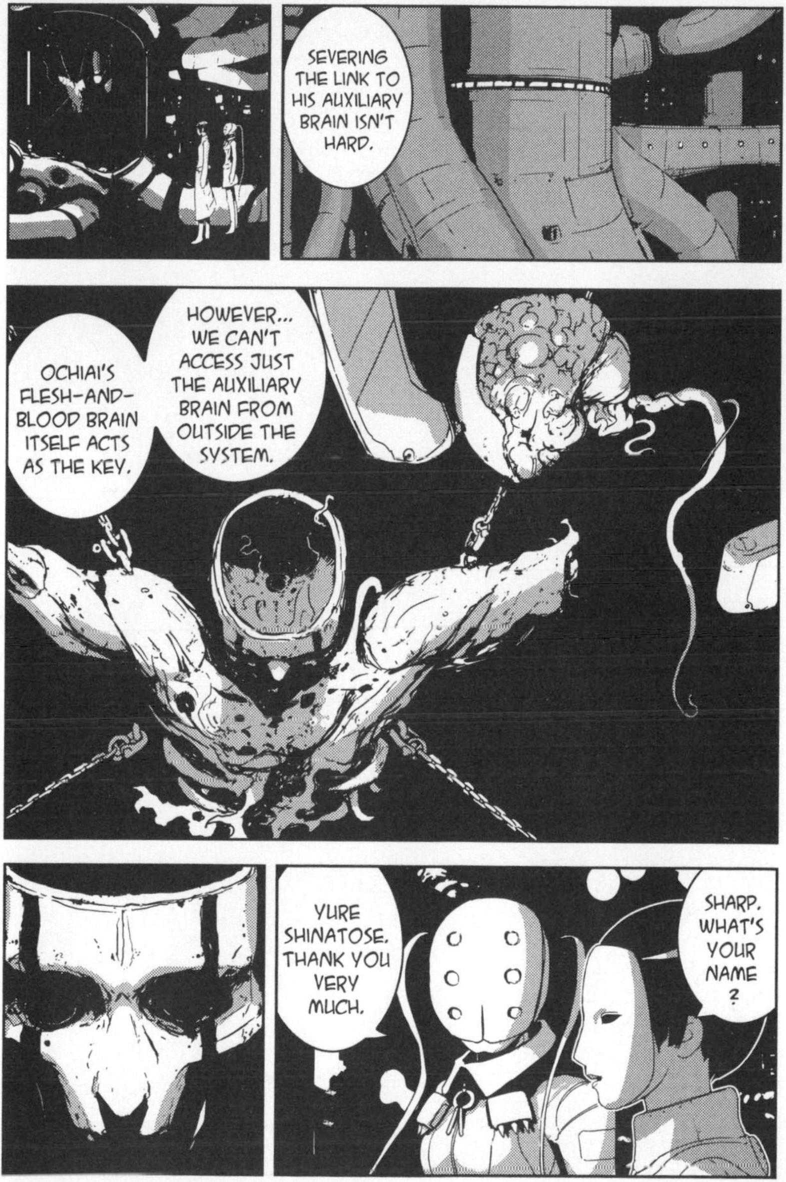
SEVERING THE LINK TO HIS AUXILIARY BRAIN ISN'T HARD.
HOWEVER... WE CAN'T ACCESS JUST THE AUXILIARY BRAIN FROM OUTSIDE THE SYSTEM.
OCHIAI'S FLESH-AND-BLOOD BRAIN ITSELF ACTS AS THE KEY.
SHARP. WHAT'S YOUR NAME?
YURE SHINATOSE. THANK YOU VERY MUCH.

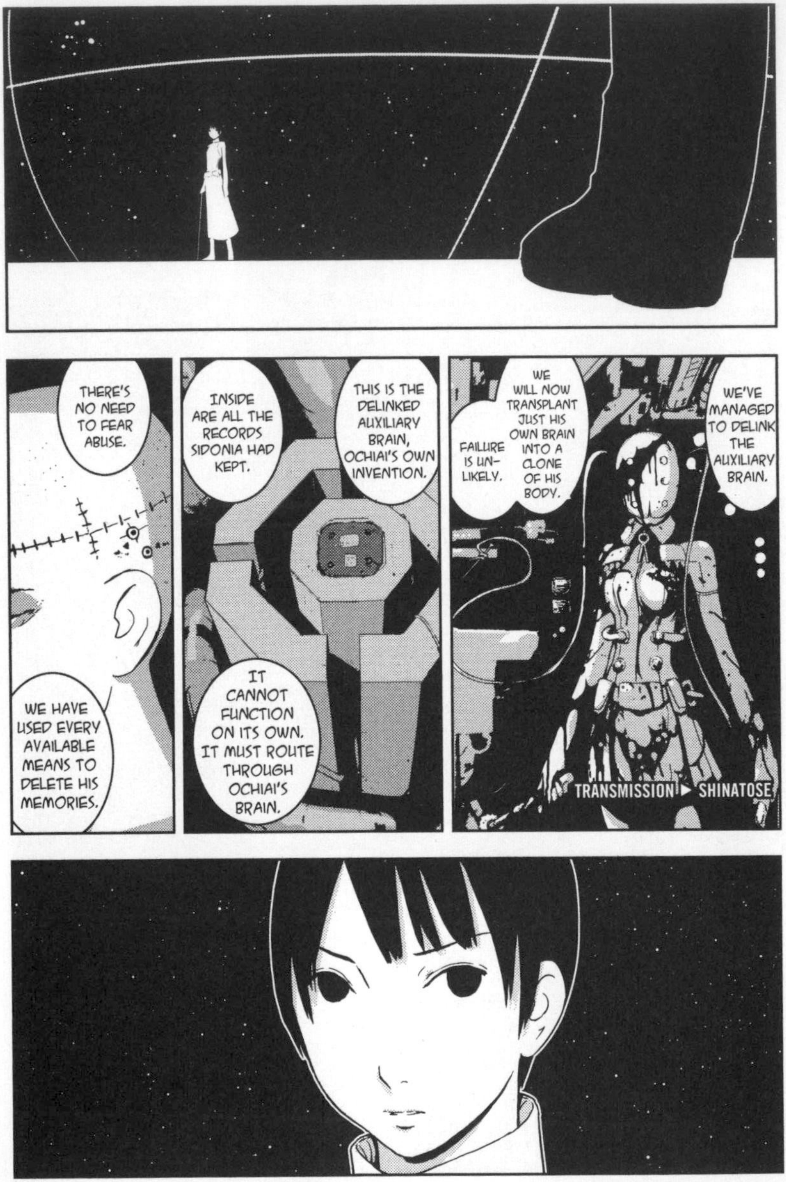
WE'VE MANAGED TO DELINK THE AUXILIARY BRAIN.
WE WILL NOW TRANSPLANT JUST HIS OWN BRAIN INTO A CLONE OF HIS BODY.
FAILURE IS UN-LIKELY.
TRANSMISSION ▶ SHINATOSE
THIS IS THE DELINKED AUXILIARY BRAIN, OCHIAI'S OWN INVENTION.
INSIDE ARE ALL THE RECORDS SIDONIA HAD KEPT.
IT CANNOT FUNCTION ON ITS OWN. IT MUST ROUTE THROUGH OCHIAI'S BRAIN.
THERE'S NO NEED TO FEAR ABUSE.
WE HAVE USED EVERY AVAILABLE MEANS TO DELETE HIS MEMORIES.

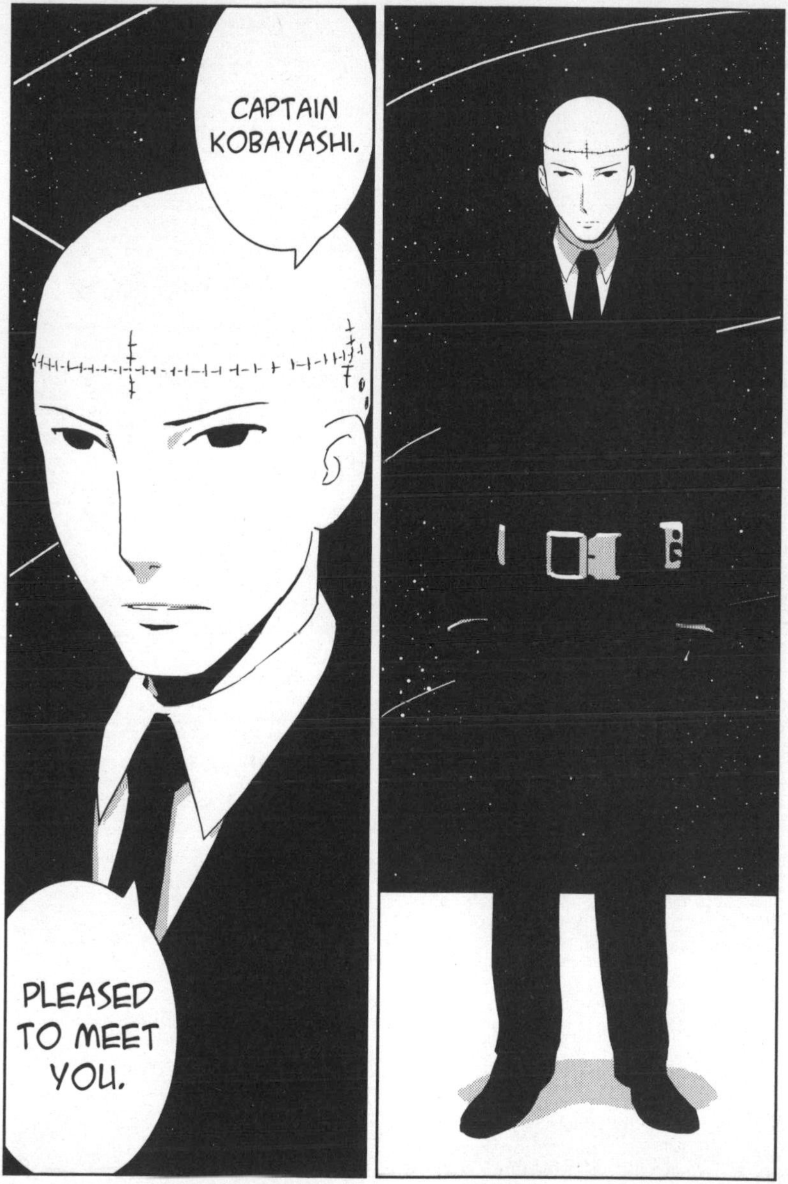
CAPTAIN KOBAYASHI.
PLEASED TO MEET YOU.

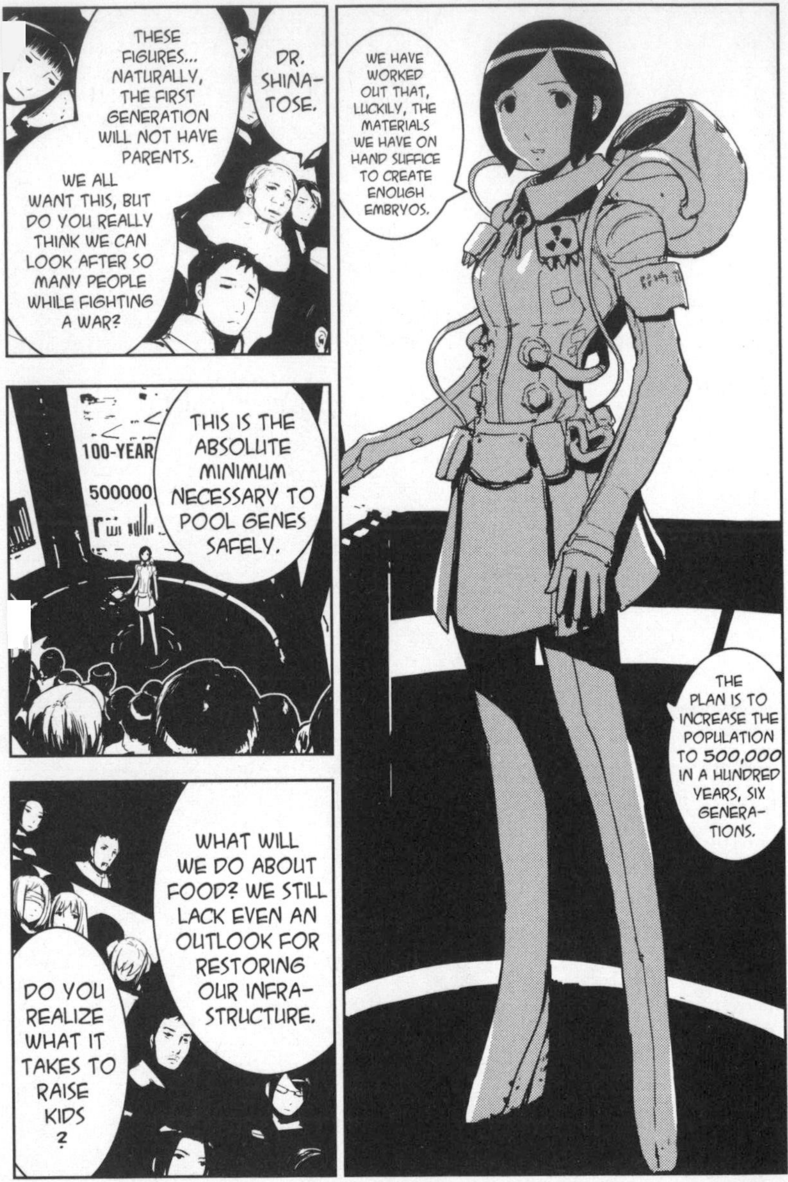
WE HAVE WORKED OUT THAT, LUCKILY, THE MATERIALS WE HAVE ON HAND SUFFICE TO CREATE ENOUGH EMBRYOS.
THE PLAN IS TO INCREASE THE POPULATION TO 500,000 IN A HUNDRED YEARS, SIX GENERA-TIONS.
DR. SHINA-TOSE.
THESE FIGURES... NATURALLY, THE FIRST GENERATION WILL NOT HAVE PARENTS.
WE ALL WANT THIS, BUT DO YOU REALLY THINK WE CAN LOOK AFTER SO MANY PEOPLE WHILE FIGHTING A WAR?
100-YEAR
500000
THIS IS THE ABSOLUTE MINIMUM NECESSARY TO POOL GENES SAFELY.
WHAT WILL WE DO ABOUT FOOD? WE STILL LACK EVEN AN OUTLOOK FOR RESTORING OUR INFRA-STRUCTURE.
DO YOU REALIZE WHAT IT TAKES TO RAISE KIDS ?

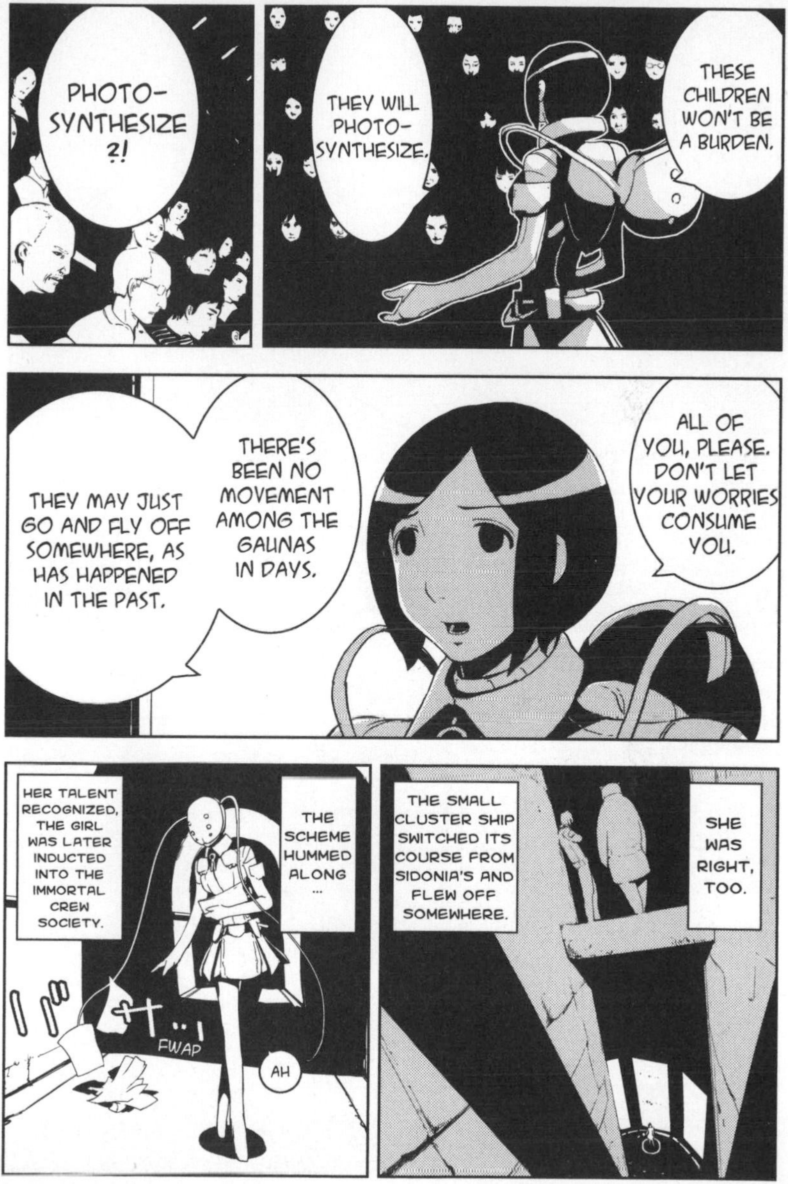
THESE CHILDREN WON'T BE A BURDEN.
THEY WILL PHOTO-SYNTHESIZE.
PHOTO-SYNTHESIZE?!
ALL OF YOU, PLEASE. DON'T LET YOUR WORRIES CONSUME YOU.
THERE'S BEEN NO MOVEMENT AMONG THE GAUNAS IN DAYS.
THEY MAY JUST GO AND FLY OFF SOMEWHERE, AS HAS HAPPENED IN THE PAST.
SHE WAS RIGHT, TOO.
THE SMALL CLUSTER SHIP SWITCHED ITS COURSE FROM SIDONIA'S AND FLEW OFF SOMEWHERE.
THE SCHEME HUMMED ALONG ...
HER TALENT RECOGNIZED, THE GIRL WAS LATER INDUCTED INTO THE IMMORTAL CREW SOCIETY.
AH
FWAP

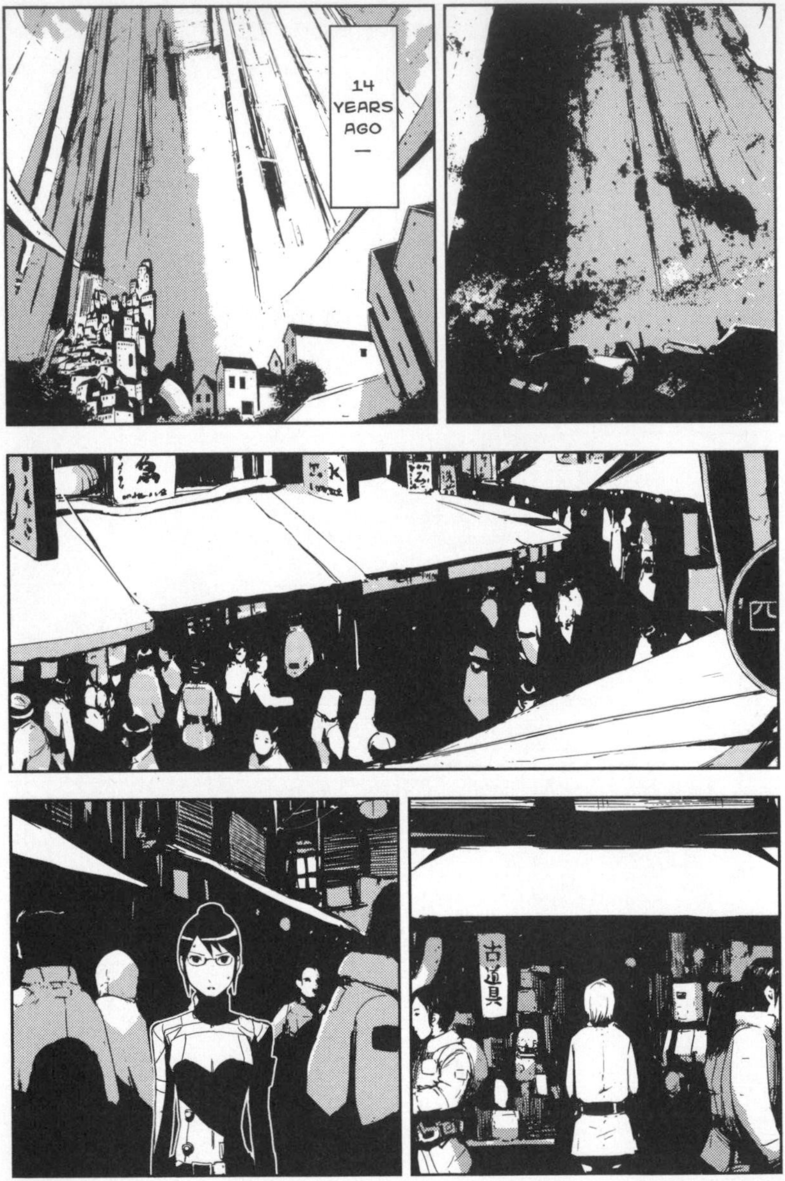
古道具
14 YEARS AGO —

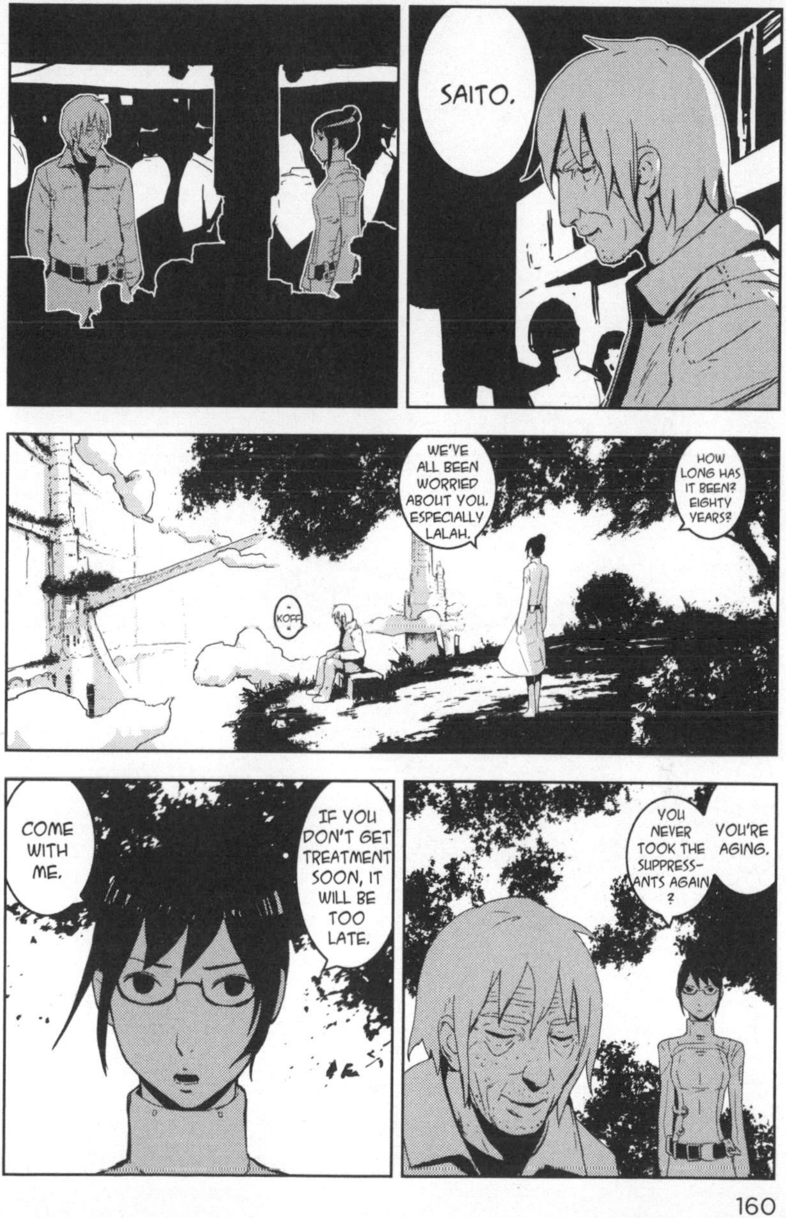
SAITO.
HOW LONG HAS IT BEEN? EIGHTY YEARS?
WE'VE ALL BEEN WORRIED ABOUT YOU. ESPECIALLY LALAH.
KOFF
YOU'RE AGING.
YOU NEVER TOOK THE SUPPRESS-ANTS AGAIN?
IF YOU DON'T GET TREATMENT SOON, IT WILL BE TOO LATE.
COME WITH ME.

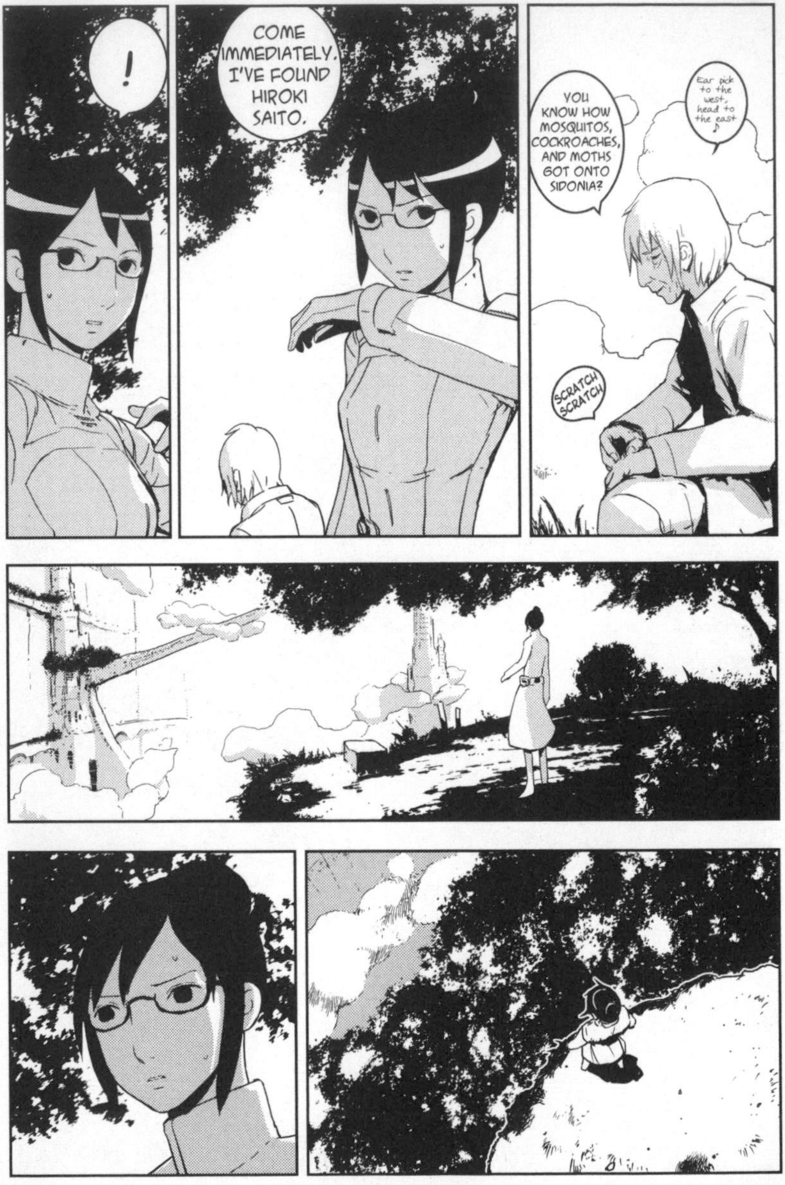
Ear pick to the west, head to the east ♪
YOU KNOW HOW MOSQUITOS, COCKROACHES, AND MOTHS GOT ONTO SIDONIA?
SCRATCH SCRATCH
COME IMMEDIATELY. I'VE FOUND HIROKI SAITO.
!

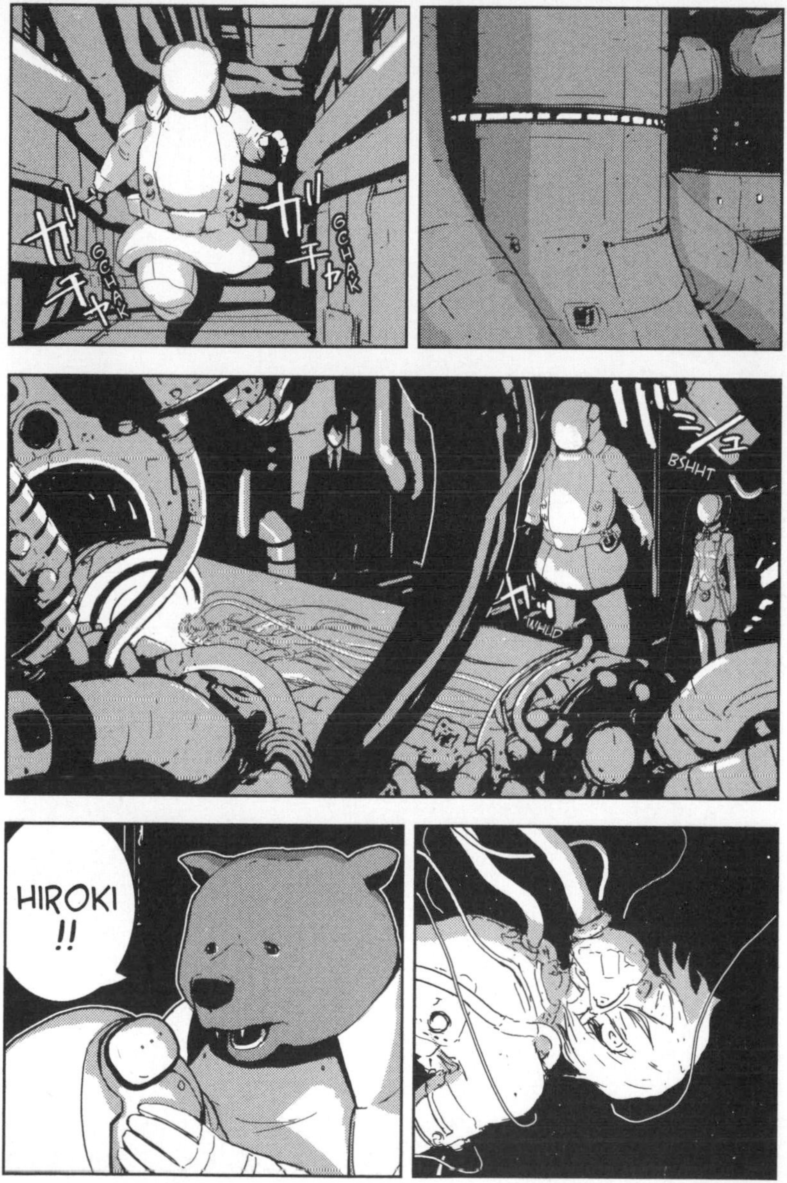
GCHAK
GCHAK
BSHHT
WHUD
HIROKI !!

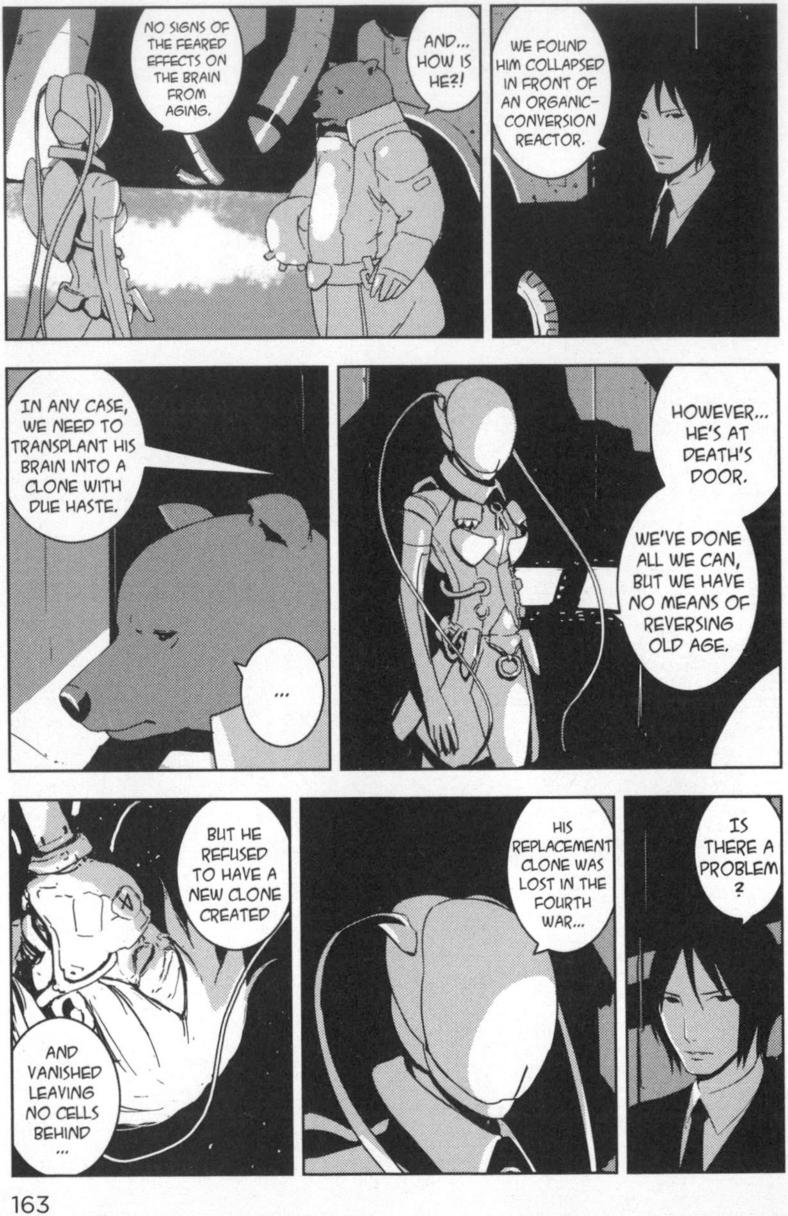
WE FOUND HIM COLLAPSED IN FRONT OF AN ORGANIC-CONVERSION REACTOR.
AND... HOW IS HE?!
NO SIGNS OF THE FEARED EFFECTS ON THE BRAIN FROM AGING.
HOWEVER... HE'S AT DEATH'S DOOR.
WE'VE DONE ALL WE CAN, BUT WE HAVE NO MEANS OF REVERSING OLD AGE.
IN ANY CASE, WE NEED TO TRANSPLANT HIS BRAIN INTO A CLONE WITH DUE HASTE.
...
IS THERE A PROBLEM?
HIS REPLACEMENT CLONE WAS LOST IN THE FOURTH WAR...
BUT HE REFUSED TO HAVE A NEW CLONE CREATED
AND VANISHED LEAVING NO CELLS BEHIND ...

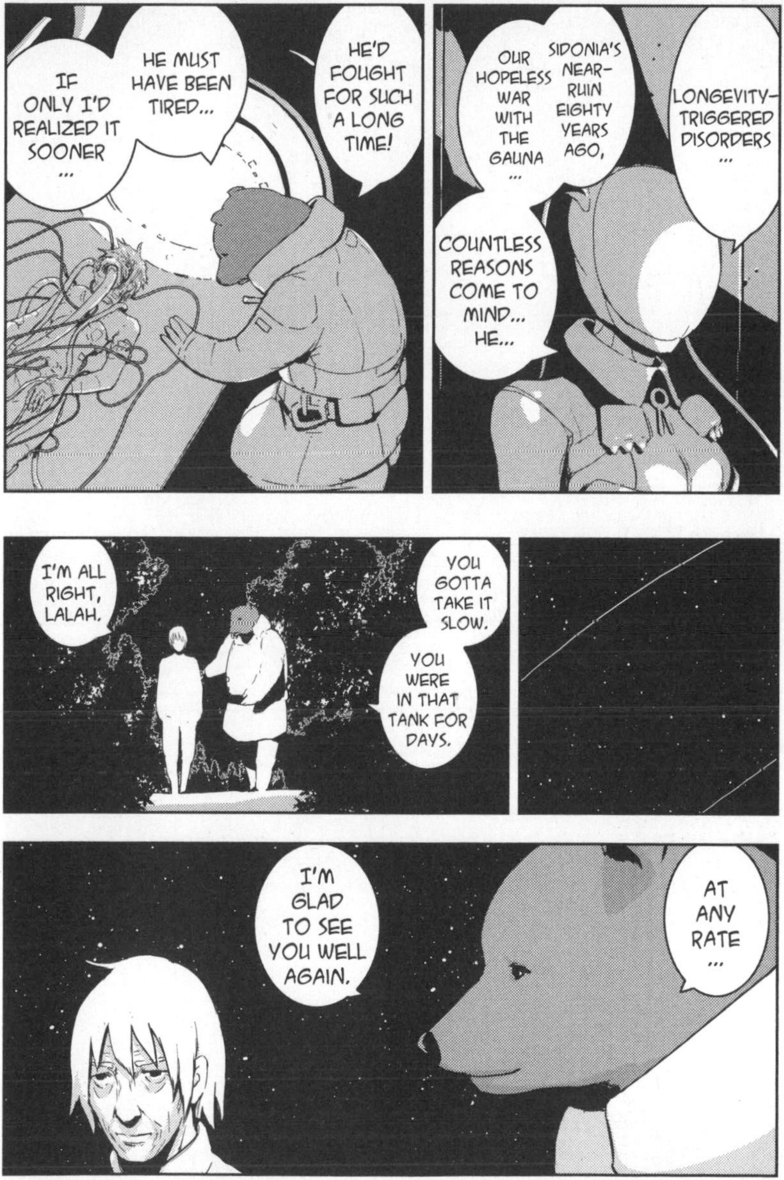
LONGEVITY-TRIGGERED DISORDERS ...
SIDONIA'S NEAR-RUIN EIGHTY YEARS AGO,
OUR HOPELESS WAR WITH THE GAUNA ...
COUNTLESS REASONS COME TO MIND... HE...
HE'D FOUGHT FOR SUCH A LONG TIME!
HE MUST HAVE BEEN TIRED...
IF ONLY I'D REALIZED IT SOONER ...
YOU GOTTA TAKE IT SLOW.
YOU WERE IN THAT TANK FOR DAYS.
I'M ALL RIGHT, LALAH.
AT ANY RATE ...
I'M GLAD TO SEE YOU WELL AGAIN.

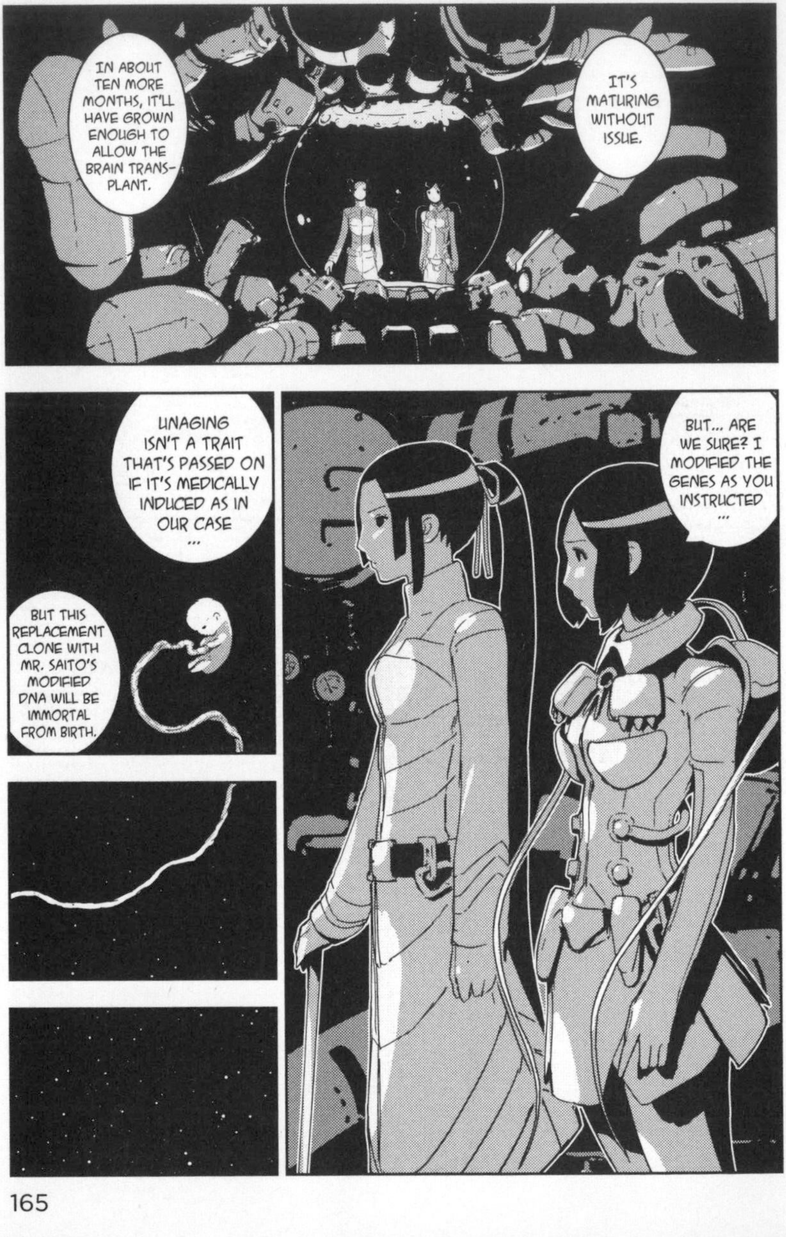
IT'S MATURING WITHOUT ISSUE.
IN ABOUT TEN MORE MONTHS, IT'LL HAVE GROWN ENOUGH TO ALLOW THE BRAIN TRANS-PLANT.
BUT... ARE WE SURE? I MODIFIED THE GENES AS YOU INSTRUCTED ...
UNAGING ISN'T A TRAIT THAT'S PASSED ON IF IT'S MEDICALLY INDUCED AS IN OUR CASE ...
BUT THIS REPLACEMENT CLONE WITH MR. SAITO'S MODIFIED DNA WILL BE IMMORTAL FROM BIRTH.

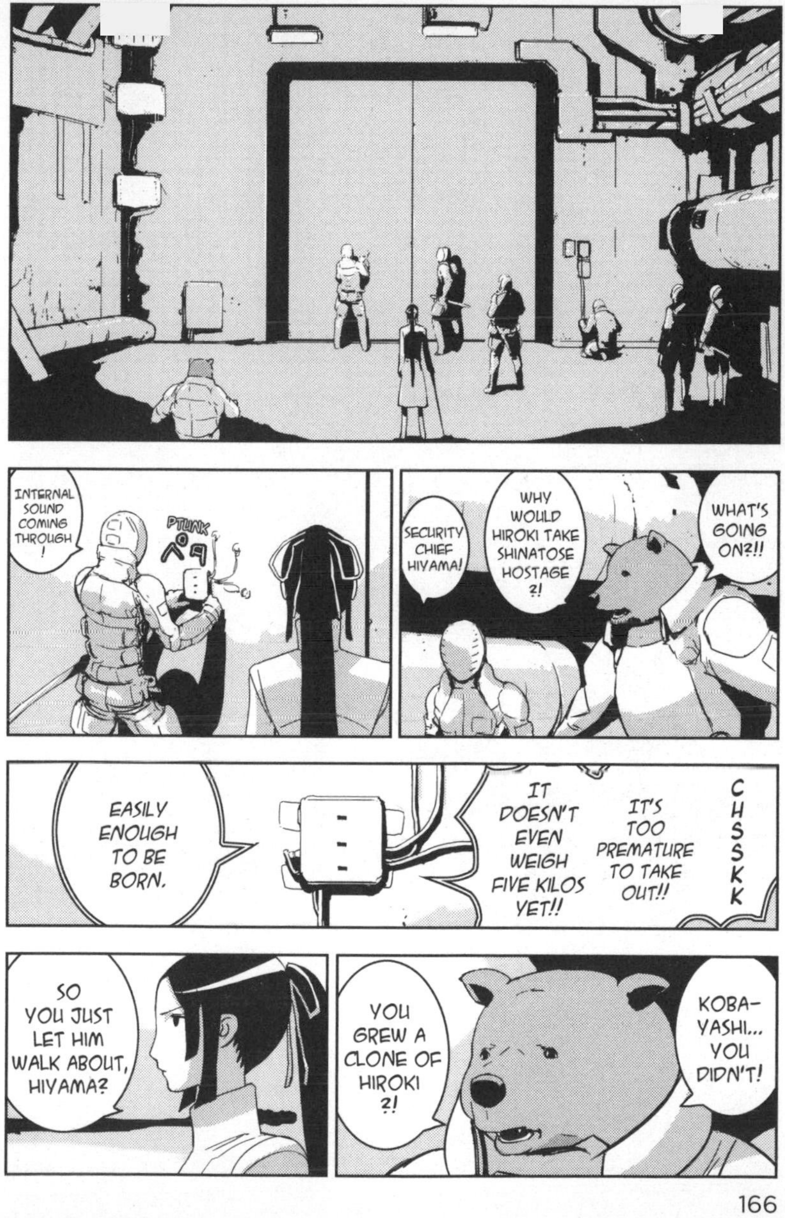
WHAT'S GOING ON?!!
WHY WOULD HIROKI TAKE SHINATOSE HOSTAGE ?!
SECURITY CHIEF HIYAMA!
INTERNAL SOUND COMING THROUGH!
PTUNK
...
CHSSKK
IT'S TOO PREMATURE TO TAKE OUT!!
IT DOESN'T EVEN WEIGH FIVE KILOS YET!!
...
EASILY ENOUGH TO BE BORN.
KOBA-YASHI... YOU DIDN'T!
YOU GREW A CLONE OF HIROKI ?!
SO YOU JUST LET HIM WALK ABOUT, HIYAMA?

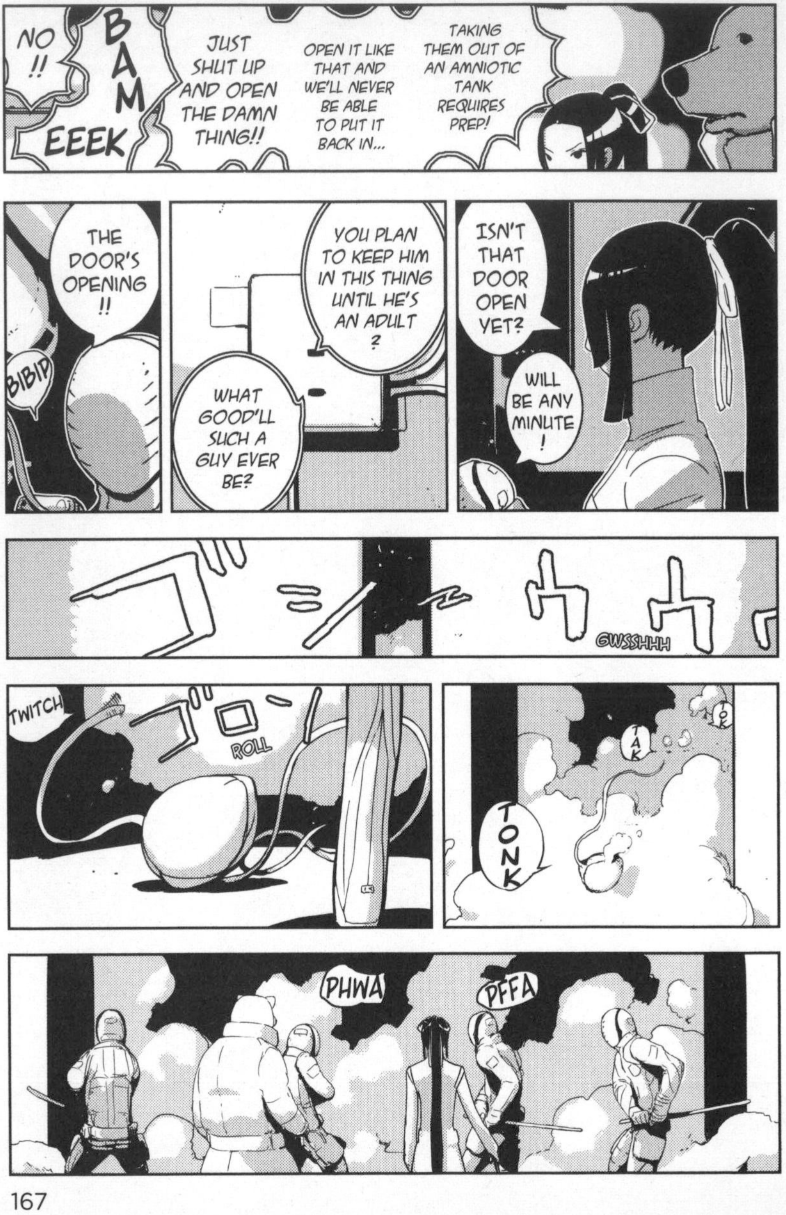
TAKING THEM OUT OF AN AMNIOTIC TANK REQUIRES PREP!
OPEN IT LIKE THAT AND WE'LL NEVER BE ABLE TO PUT IT BACK IN...
JUST SHUT UP AND OPEN THE DAMN THING!!
BAM
NO !!
EEEK
ISN'T THAT DOOR OPEN YET?
WILL BE ANY MINUTE !
YOU PLAN TO KEEP HIM IN THIS THING UNTIL HE'S AN ADULT ?
WHAT GOOD'LL SUCH A GUY EVER BE?
THE DOOR'S OPENING !!
BIBIP
GWSSHHH
TOK
TAK
TONK
TWITCH
ROLL
PFFA
PHWA

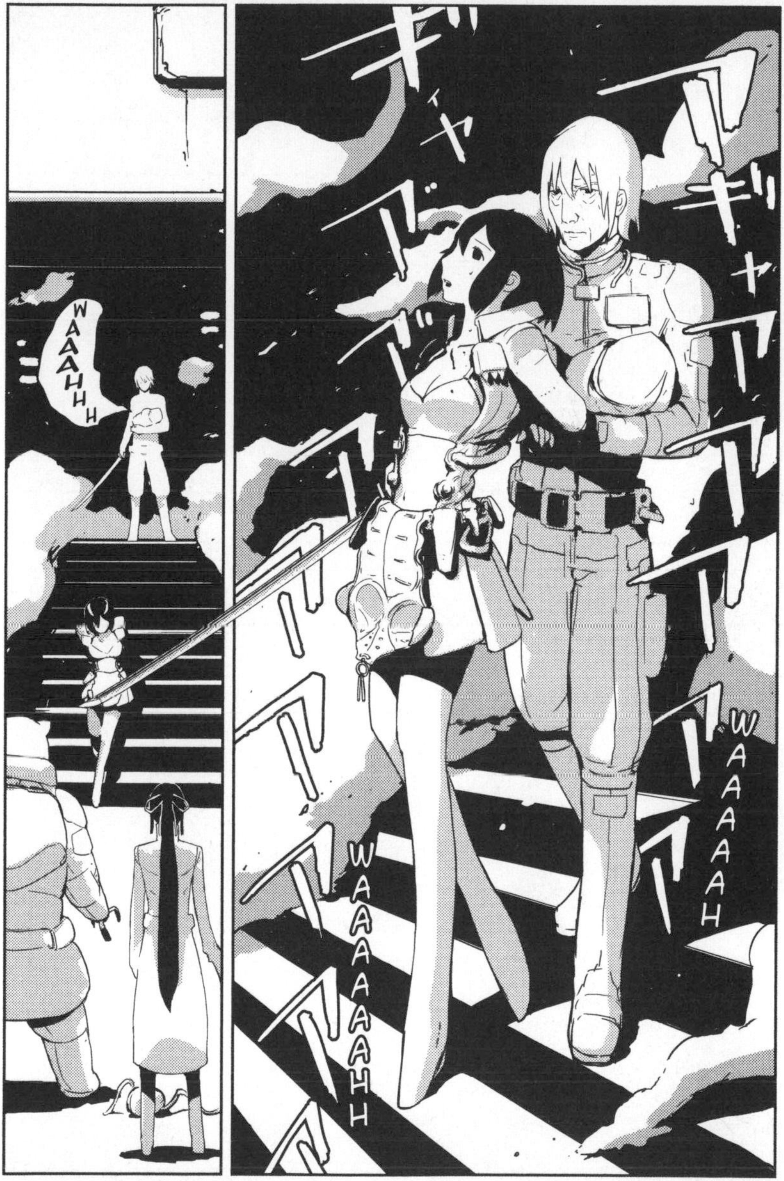
WAAAAAH
WAAAAAAHH
WAAAHHH

HIROKI !!!
YOU TOOK THE CLONE OUT OF THE AMNIOTIC TANK?!
HAVE YOU FORGOTTEN THE OATH OF THE SOCIETY ?
YOUR LIFE AND DEATH ARE NO LONGER YOUR OWN CALL.
YOUR EXPERIENCE AND KNOWLEDGE BELONG TO THE SHIP.
GIVE ME FIFTEEN YEARS... NO, TEN WILL DO.
WAAHH
TWITCH
I SWEAR I'LL TEACH THIS CHILD EVERYTHING I KNOW.
WAAAHH

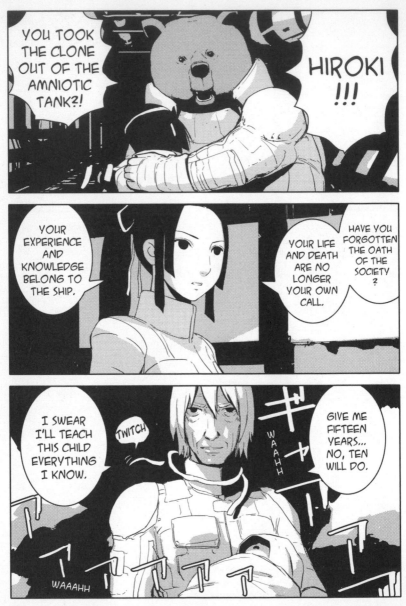

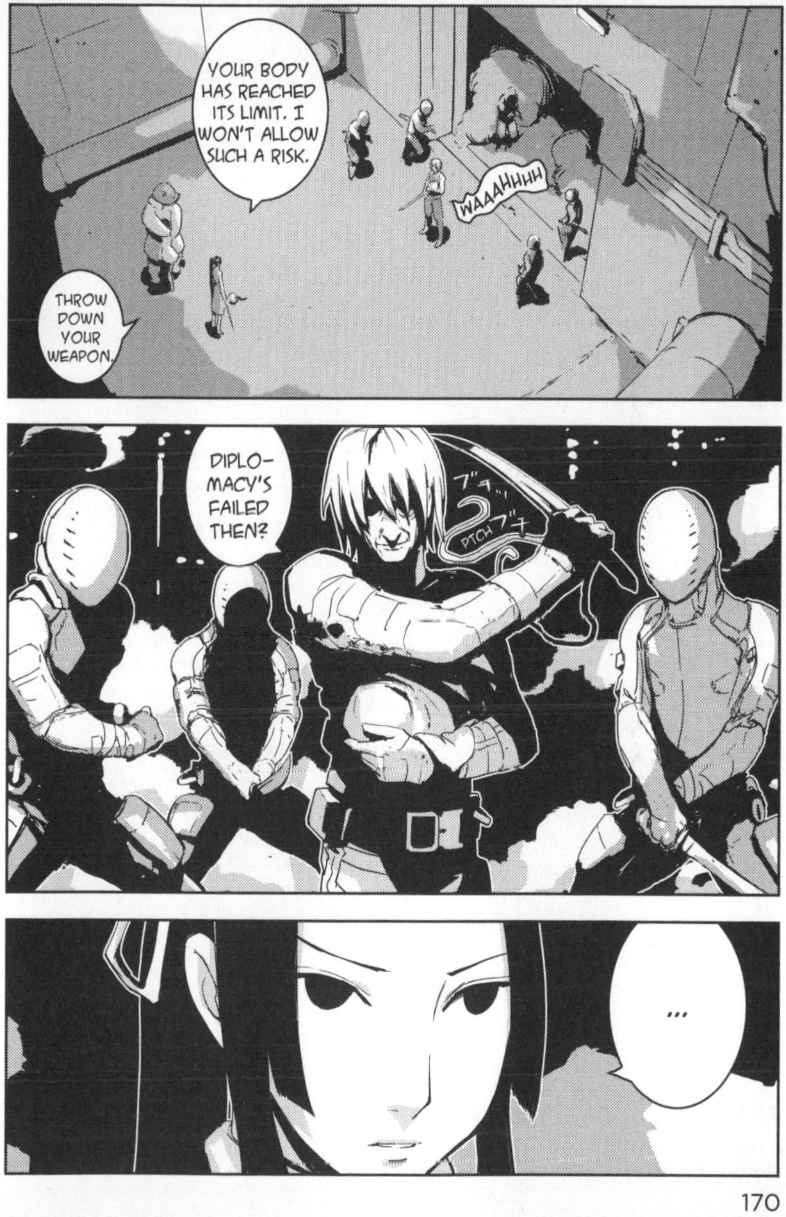
YOUR BODY HAS REACHED ITS LIMIT. I WON'T ALLOW SUCH A RISK.
WAAAHHHH
THROW DOWN YOUR WEAPON.
PTCH
DIPLO-MACY'S FAILED THEN?
...

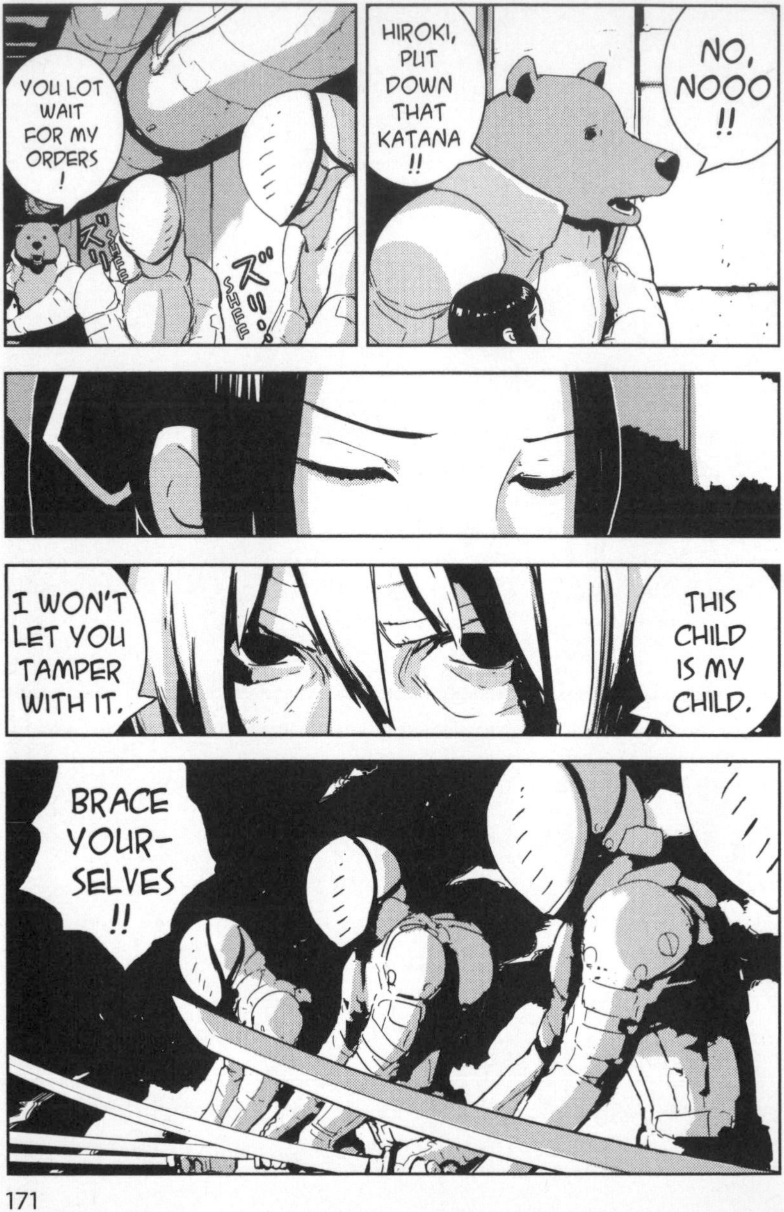
NO, NOOO !!
HIROKI, PUT DOWN THAT KATANA !!
YOU LOT WAIT FOR MY ORDERS !
SHEE
SHEFF
THIS CHILD IS MY CHILD.
I WON'T LET YOU TAMPER WITH IT.
BRACE YOUR-SELVES !!

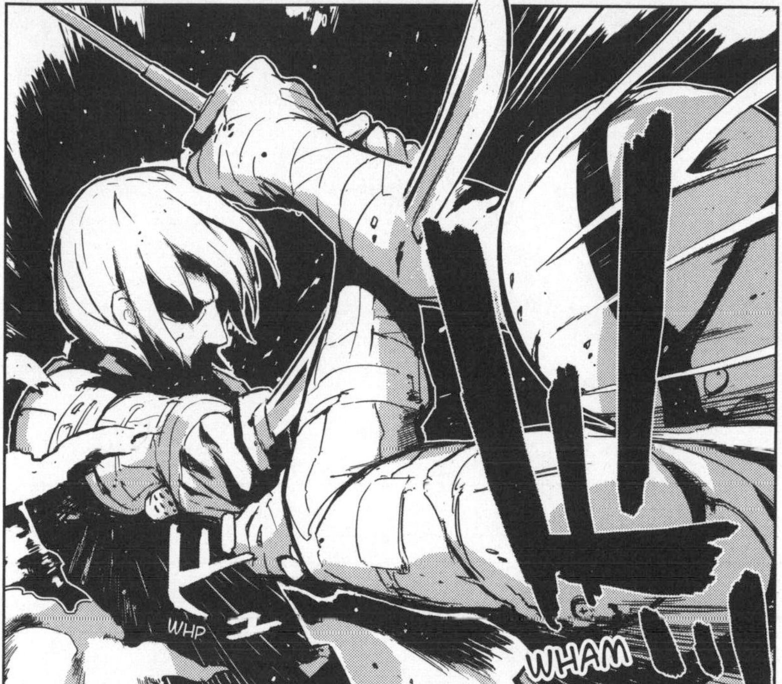
WHP
WHAM

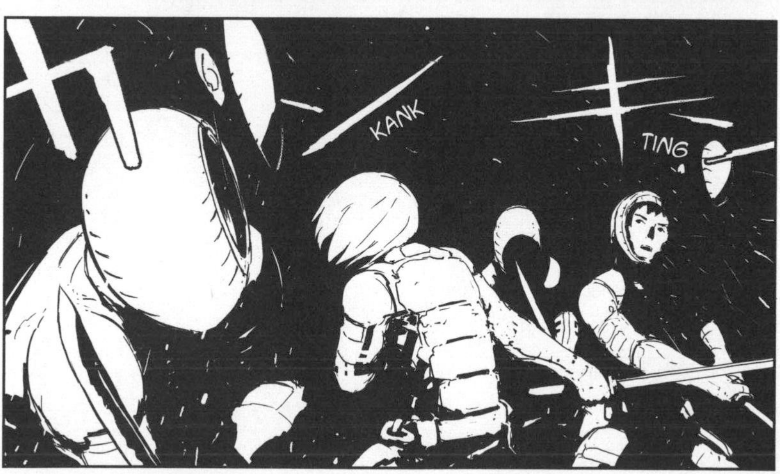
KANK
TING

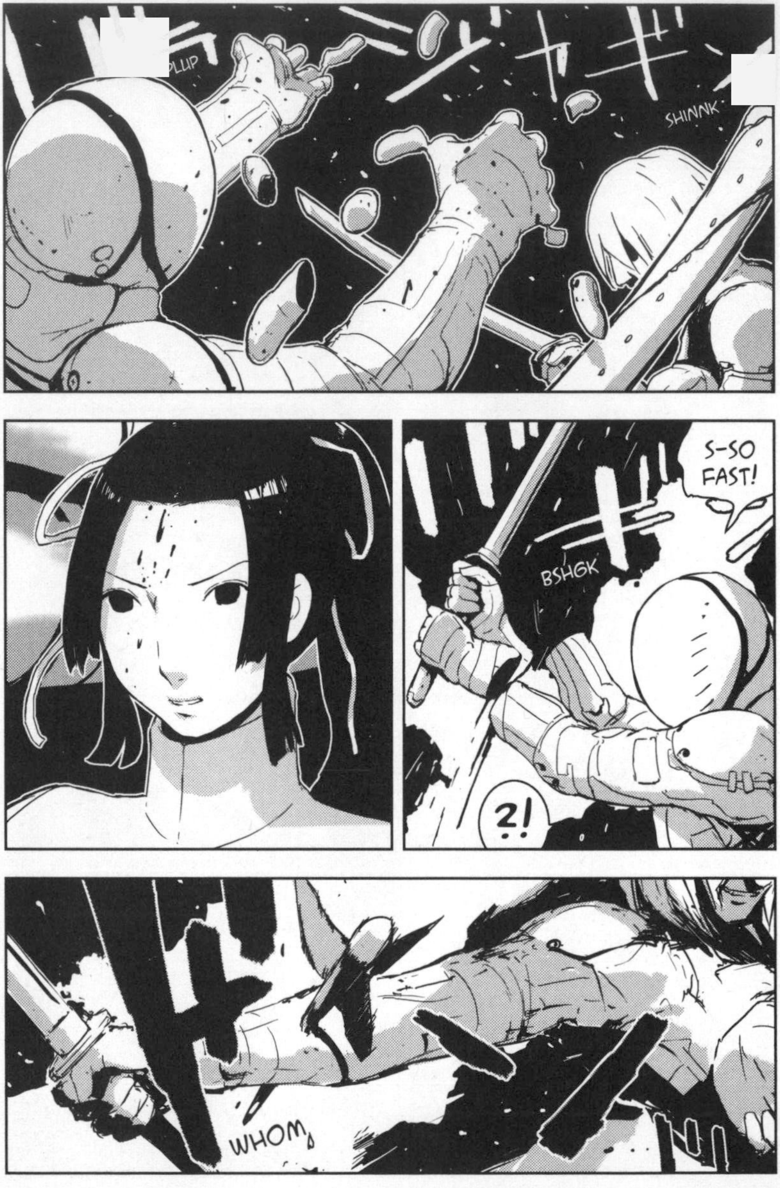

SHINNK
PLUP
S-SO FAST!
BSHGK
?!
WHOM

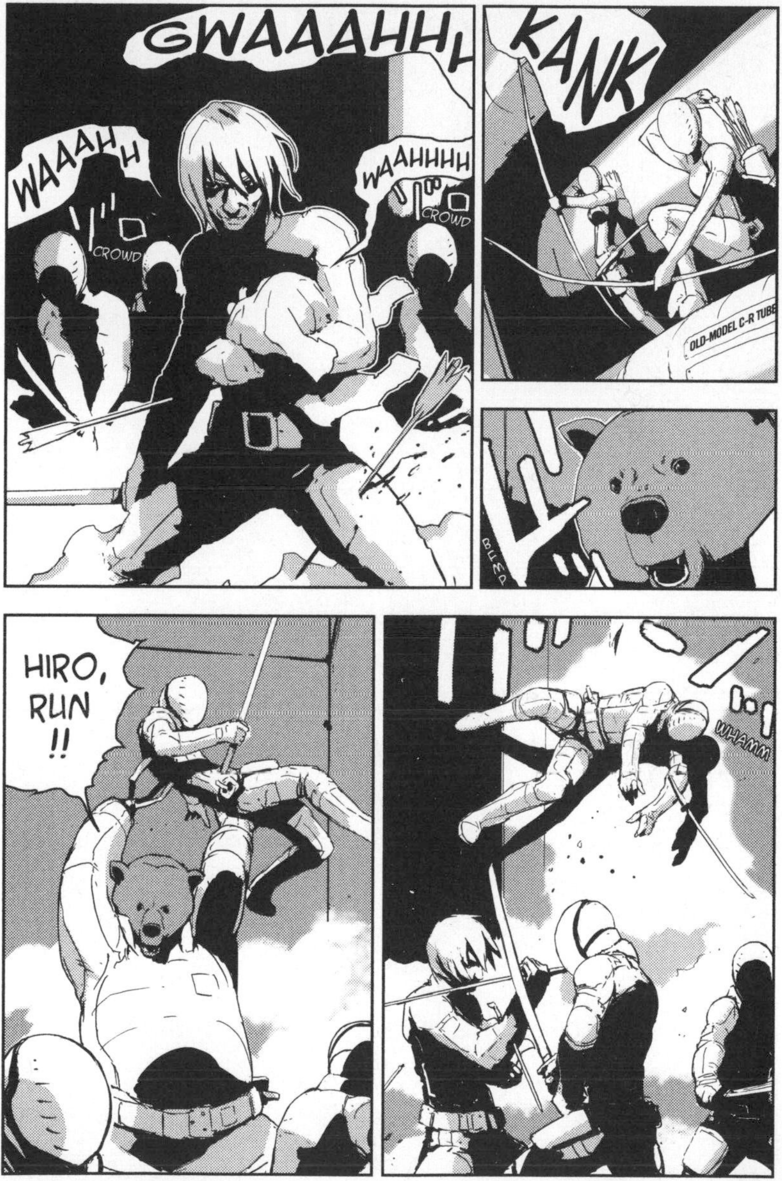
KANK
OLD-MODEL C-R TUBE
BEMP
GWAAAHHH
WAAHHHH
CROWD
WAAAHH
CROWD
WHAMM
HIRO, RUN !!

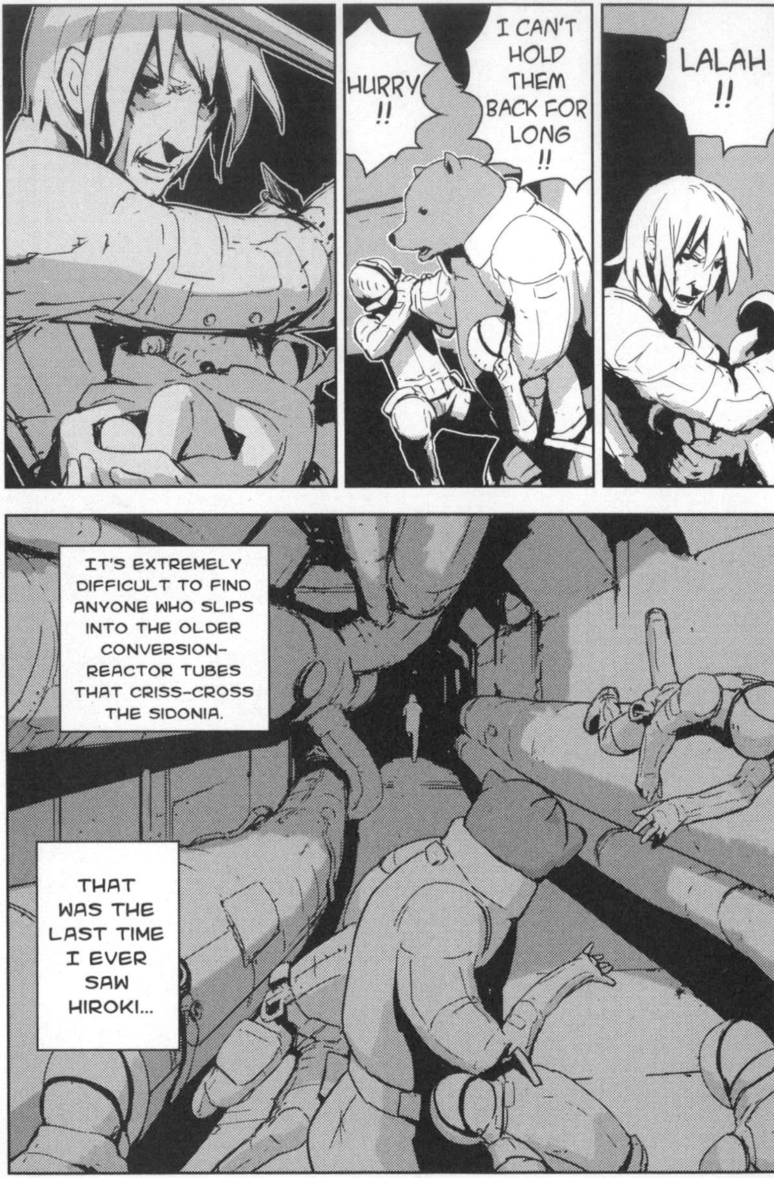
LALAH !!
I CAN'T HOLD THEM BACK FOR LONG !!
HURRY !!
IT'S EXTREMELY DIFFICULT TO FIND ANYONE WHO SLIPS INTO THE OLDER CONVERSION-REACTOR TUBES THAT CRISS-CROSS THE SIDONIA.
THAT WAS THE LAST TIME I EVER SAW HIROKI...

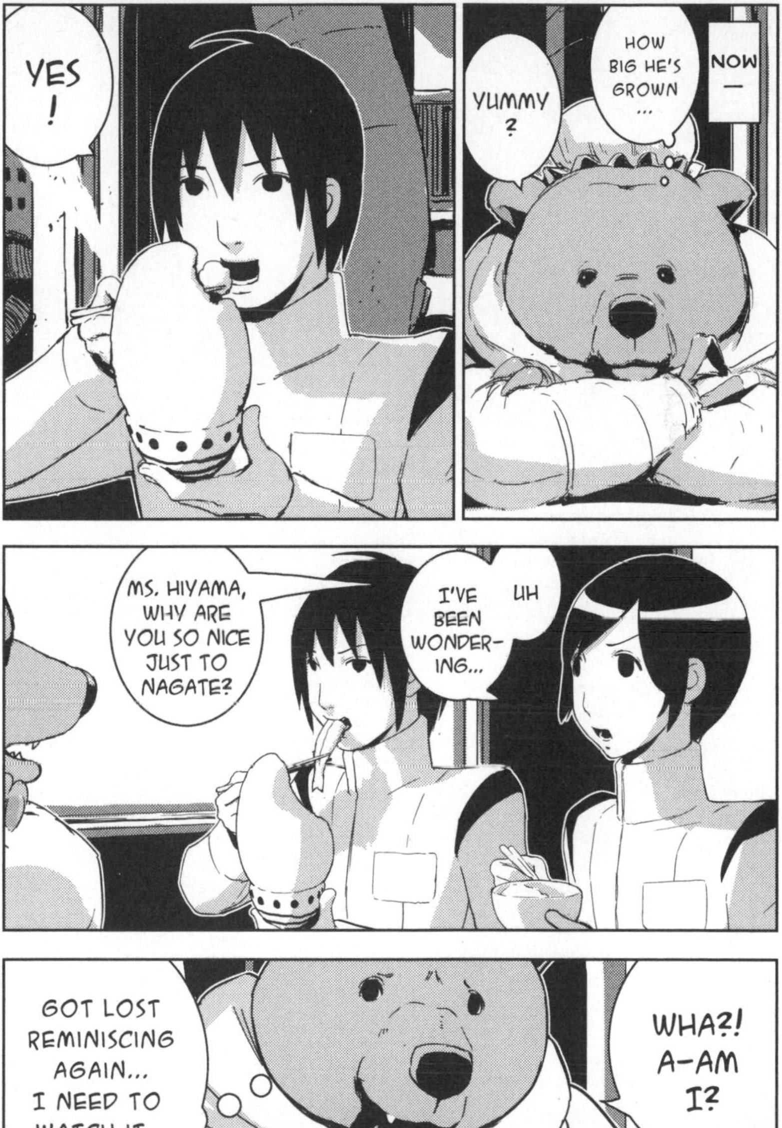
NOW —
HOW BIG HE'S GROWN ...
YUMMY?
YES!
UH
I'VE BEEN WONDERING...
MS. HIYAMA, WHY ARE YOU SO NICE JUST TO NAGATE?
WHA?! A-AM I?
GOT LOST REMINISCING AGAIN... I NEED TO WATCH IT...

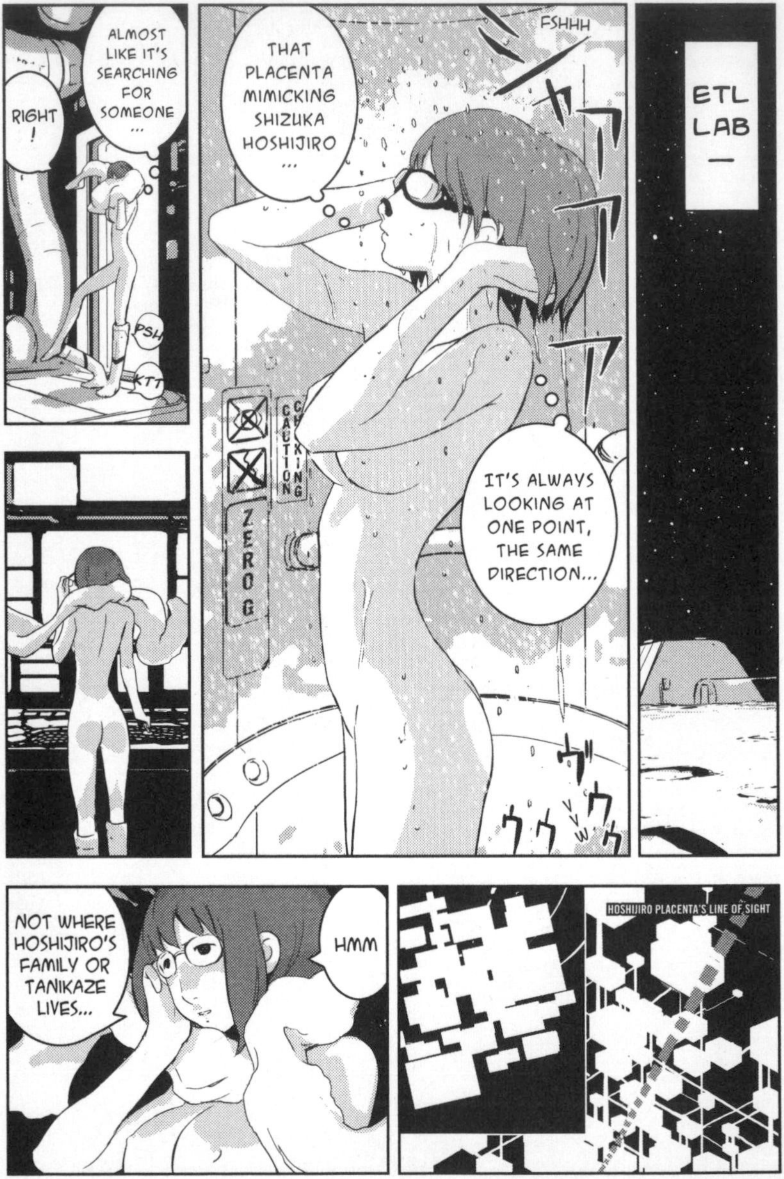
ETL LAB —
FSHHH
THAT PLACENTA MIMICKING SHIZUKA HOSHIJIRO ...
CAUTION
CHECKING
ZEROG
IT'S ALWAYS LOOKING AT ONE POINT, THE SAME DIRECTION...
ALMOST LIKE IT'S SEARCHING FOR SOMEONE ...
RIGHT !
PSH
KTT
HOSHIJIRO PLACENTA'S LINE OF SIGHT
HMM
NOT WHERE HOSHIJIRO'S FAMILY OR TANIKAZE LIVES...

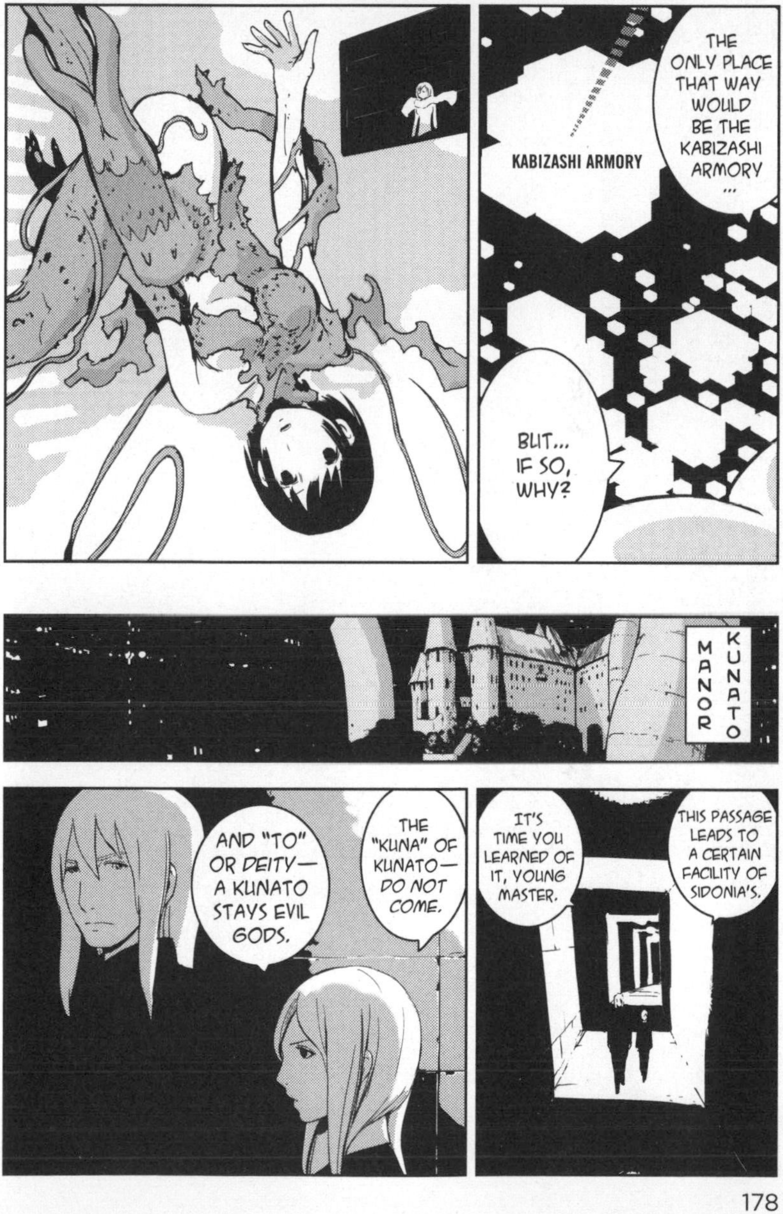
THE ONLY PLACE THAT WAY WOULD BE THE KABIZASHI ARMORY ...
KABIZASHI ARMORY
BUT... IF SO, WHY?
KUNATO MANOR
THIS PASSAGE LEADS TO A CERTAIN FACILITY OF SIDONIA'S.
IT'S TIME YOU LEARNED OF IT, YOUNG MASTER.
THE "KUNA" OF KUNATO— DO NOT COME.
AND "TO" OR DEITY— A KUNATO STAYS EVIL GODS.

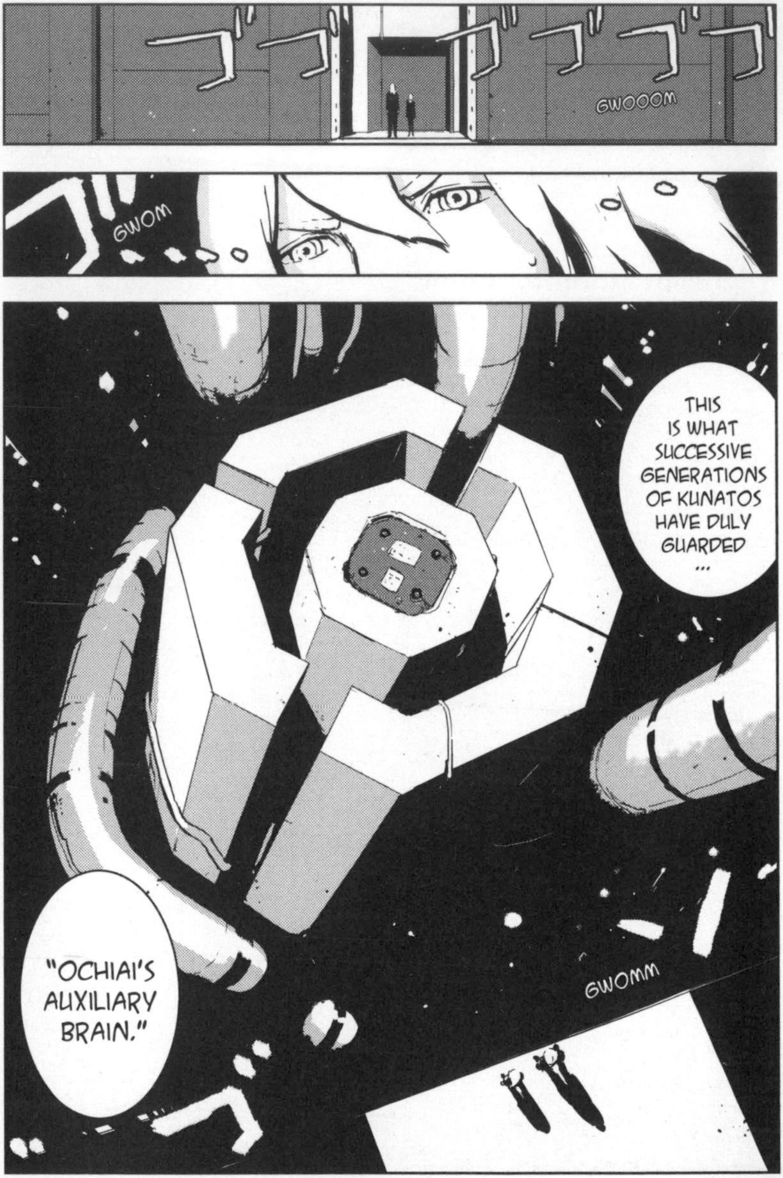
GWOOOM
GWOM
THIS IS WHAT SUCCESSIVE GENERATIONS OF KUNATOS HAVE DULY GUARDED ...
"OCHIAI'S AUXILIARY BRAIN."
GWOMM

KNIGHTS OF SIDONIA Volume③: END

Knights of Sidonia, volume 3

Translation: Kumar Sivasubramanian
Production: Grace Lu
Daniela Yamada
Tomoe Tsutsumi

First published in Japan in 2010 by Kodansha, Ltd., Tokyo
Publication for this English edition arranged through Kodansha, Ltd., Tokyo
English language version produced by Vertical, Inc.

Translation provided by Vertical, Inc., 2013
Published by Vertical, Inc., New York

Originally published in Japanese as *Shidonia no Kishi 3* by Kodansha, Ltd.
Shidonia no Kishi first serialized in *Afternoon*, Kodansha, Ltd., 2009-

ISBN: 978-1-935654-82-7

Manufactured in Canada

First Edition

Second Printing

Vertical, Inc.
451 Park Avenue South
7th Floor
New York, NY 10016
www.vertical-inc.com